Communications
in Computer and Information Science 2125

Series Editors

Gang Li ⓘ, *School of Information Technology, Deakin University, Burwood, VIC, Australia*
Joaquim Filipe ⓘ, *Polytechnic Institute of Setúbal, Setúbal, Portugal*
Zhiwei Xu, *Chinese Academy of Sciences, Beijing, China*

Rationale
The CCIS series is devoted to the publication of proceedings of computer science conferences. Its aim is to efficiently disseminate original research results in informatics in printed and electronic form. While the focus is on publication of peer-reviewed full papers presenting mature work, inclusion of reviewed short papers reporting on work in progress is welcome, too. Besides globally relevant meetings with internationally representative program committees guaranteeing a strict peer-reviewing and paper selection process, conferences run by societies or of high regional or national relevance are also considered for publication.

Topics
The topical scope of CCIS spans the entire spectrum of informatics ranging from foundational topics in the theory of computing to information and communications science and technology and a broad variety of interdisciplinary application fields.

Information for Volume Editors and Authors
Publication in CCIS is free of charge. No royalties are paid, however, we offer registered conference participants temporary free access to the online version of the conference proceedings on SpringerLink (http://link.springer.com) by means of an http referrer from the conference website and/or a number of complimentary printed copies, as specified in the official acceptance email of the event.

CCIS proceedings can be published in time for distribution at conferences or as postproceedings, and delivered in the form of printed books and/or electronically as USBs and/or e-content licenses for accessing proceedings at SpringerLink. Furthermore, CCIS proceedings are included in the CCIS electronic book series hosted in the SpringerLink digital library at http://link.springer.com/bookseries/7899. Conferences publishing in CCIS are allowed to use Online Conference Service (OCS) for managing the whole proceedings lifecycle (from submission and reviewing to preparing for publication) free of charge.

Publication process
The language of publication is exclusively English. Authors publishing in CCIS have to sign the Springer CCIS copyright transfer form, however, they are free to use their material published in CCIS for substantially changed, more elaborate subsequent publications elsewhere. For the preparation of the camera-ready papers/files, authors have to strictly adhere to the Springer CCIS Authors' Instructions and are strongly encouraged to use the CCIS LaTeX style files or templates.

Abstracting/Indexing
CCIS is abstracted/indexed in DBLP, Google Scholar, EI-Compendex, Mathematical Reviews, SCImago, Scopus. CCIS volumes are also submitted for the inclusion in ISI Proceedings.

How to start
To start the evaluation of your proposal for inclusion in the CCIS series, please send an e-mail to ccis@springer.com.

Manisha Malhotra
Editor

Innovation and Emerging Trends in Computing and Information Technologies

First International Conference, IETCIT 2024
Mohali, Punjab, India, March 1–2, 2024
Proceedings, Part I

Editor
Manisha Malhotra ⓘ
Chandigarh University
Mohali, Punjab, India

ISSN 1865-0929   ISSN 1865-0937 (electronic)
Communications in Computer and Information Science
ISBN 978-3-031-80838-8   ISBN 978-3-031-80839-5 (eBook)
https://doi.org/10.1007/978-3-031-80839-5

© The Editor(s) (if applicable) and The Author(s), under exclusive license
to Springer Nature Switzerland AG 2025

This work is subject to copyright. All rights are solely and exclusively licensed by the Publisher, whether the whole or part of the material is concerned, specifically the rights of translation, reprinting, reuse of illustrations, recitation, broadcasting, reproduction on microfilms or in any other physical way, and transmission or information storage and retrieval, electronic adaptation, computer software, or by similar or dissimilar methodology now known or hereafter developed.
The use of general descriptive names, registered names, trademarks, service marks, etc. in this publication does not imply, even in the absence of a specific statement, that such names are exempt from the relevant protective laws and regulations and therefore free for general use.
The publisher, the authors and the editors are safe to assume that the advice and information in this book are believed to be true and accurate at the date of publication. Neither the publisher nor the authors or the editors give a warranty, expressed or implied, with respect to the material contained herein or for any errors or omissions that may have been made. The publisher remains neutral with regard to jurisdictional claims in published maps and institutional affiliations.

This Springer imprint is published by the registered company Springer Nature Switzerland AG
The registered company address is: Gewerbestrasse 11, 6330 Cham, Switzerland

If disposing of this product, please recycle the paper.

To all the visionary scientists, researchers and trailblazers who are a driving force in innovation and shaping the future of computing and Information Technology. Your creativity, passion and dedication inspire us all to explore the new power of data for the betterment of society. This book is dedicated to you.

Foreword

Innovation is the stimulus of computing and the Information Technology Industry. In the fast-evolving landscape of data science, it is crucial for professionals to stay abreast of the latest trends and development. These proceedings volumes offer valuable insights into the cutting-edge technologies shaping the future of computing and Information Technology.

The books cover a wide range of topics from Artificial Intelligence, Machine Learning, Internet of Things and Data Science. The authors have done an excellent job on complex and topical problems. All the problems have been explored and addressed in an expanded manner. These books will serve as a valuable resource for all learners and readers who are looking to stay ahead in the world of computing.

Preface

The rapid pace of technological evolution has transformed the way we live, work, and interact with each other. Computing and information technology play a pivotal role in driving these changes, offering unprecedented opportunities and challenges across various sectors. From artificial intelligence and machine learning to cloud computing, cybersecurity, and beyond, the landscape of computing and information technology is continuously evolving, presenting new avenues for innovation and exploration.

It is with great pleasure that we present you the proceedings of the International Conference on Innovation & Emerging Trends in Computing & Information Technology (IETCIT 2024). This conference served as a vital platform for scholars, researchers, professionals, and enthusiasts to convene, exchange ideas, and explore the latest advancements in the field of computing and information technology. This year, the conference received a total of 417 submissions, setting a strong precedent for future years. After the peer-review process, 44 papers were selected for publication. Each paper received three reviews. Expert reviewers assessed the manuscripts for their originality, methodological soundness, and potential for impact on the field. Authors received detailed feedback to refine their work, fostering a collaborative and constructive evaluation environment.

With this conference, we aimed to provide a forum for dialogue and collaboration, fostering interdisciplinary discussions that transcend traditional boundaries and inspire ground-breaking research. Through keynote speeches, paper presentations, workshops, and networking sessions, participants had the opportunity to engage with leading experts, share their insights, and contribute to the collective understanding of emerging trends and innovative practices in computing and information technology.

As we navigate the complexities of a digital age marked by rapid change and disruption, it is essential to stay informed, adaptable, and proactive in harnessing the power of technology for the greater good. This conference served as a catalyst for such endeavours, empowering attendees to drive positive change and shape the future of computing and information technology.

November 2024   Manisha Malhotra

Acknowledgment

We hope that the International Conference on Innovation & Emerging Trends in Computing & Information Technology was a memorable and enriching experience for all, and we hope you enjoyed fruitful discussions, meaningful connections, and impactful outcomes.

We extend our heartfelt gratitude to all the participants, sponsors, organizers, and volunteers who contributed to making this conference a reality. Your dedication and support were invaluable, and we are confident that this event served as a catalyst for transformative ideas and collaborations that will propel the field of computing and information technology forward.

We wish to acknowledge the valuable contributions of the reviewers who reviewed the submissions and gave their valuable inputs to enhance the quality of the papers.

We are very grateful to Communications in Computer and Information Science (CCIS) of Springer for this opportunity and for their belief in us. We extend our heartiest thanks and appreciate to the team for providing constant technical support.

Last but not least we are thankful to our readers for choosing these conference proceedings. We hope it will be an abundant resource for you.

"Some people don't like change, but you need to embrace change if the alternative is disaster."

Elon Musk

Organization

General Chair

Manisha Malhotra   Chandigarh University, India

Program Committee Chairs

Amanpreet Kaur Sandhu	Chandigarh University, India
Abdullah	Chandigarh University, India
Kavita Gupta	Chandigarh University, India

Steering Committee

Raj Kumar Buyya	University of Melbourne, Australia
Gurpreet Singh	Chandigarh University, India
Samia Chehbi Gamoura	University of Strasbourg, France
Shadi Aljawarneh	Jordan University of Science and Technology, Jordan
Disha Handa	Chandigarh University, India
Krishan Tuli	Chandigarh University, India
Shivani Jaswal	National College of Ireland, Ireland
Kawaljit Kaur	Chandigarh University, India
Sarabjeet Kaur	Chandigarh University, India
Mohammad Zunnun Khan	University of Bisha, KSA
Pooja Thakur	Chandigarh University, India
Keshav Kumar	Chandigarh University, India

Program Committee

Luc Arnal	University of Geneva, Switzerland
Mohammad Zubair Khan	Taibah University, KSA
Mohammad Shahid Husain	University of Technology and Applied Sciences, Oman
Rajanarayanan	Arba Minch University, Ethiopia
Ravi Shankar Shukla	Saudi Electronic University, KSA

D. M. A. Hussain	Aalborg University, Denmark
Sri Chusri Haryanti	Universitas YARSI, Indonesia
Rahul Shukla	Walter Sisulu University, South Africa
H. M. Dipu Kabir	Deakin University, Australia
Vandna Jowaheer	University of Mauritius, Mauritius
Manoj Sharma	Bennett University, India
Nilanjan Dey	Techno International New Town, India
R. K. Singh	Dr. Shakuntala Mishra National Rehabilitation University, India
Ajay Kumar	Thapar Institute of Engineering and Technology, India
Sarabjeet Singh	Panjab University, India
Jarnail Singh	Presidency University, India
Siva Balan	Christ University, India
Ajay Prasad	UPES, India
Anita Gehlot	Uttaranchal University, India
Biswajit Pandey	Jain University, India
Kanwal Garg	Kurukshetra University, India
Vishal Goyal	Poornima University, India
Sarvesh Kumar	Indian Institute of Space Science and Technology, India
Yashveer Yadav	Stellantis, India
Anupam Yadav	Dr. B. R. Ambedkar National Institute of Technology, India

Additional Reviewers

Vikram Singh	Prabhash Pathak
Anupam Bhatia	Priyank Singhal
Kanwal Garg	Sandhya Bansal
Gaurav Gupta	Manish Kumar Soni
Muskan Kumari	Singara Singh
Anu Chawla	Gurjit Singh Bhathal
T. P. Singh	R. K. Singh
Ajay Kumar	Yashveer Yadav
Anil Kumar Verma	Arunesh Garg
Ravinder Kumar	Ajay Prasad
Bishwajeet Pandey	Alok Agarwal
Manoj Sharma	H. M. Dipu Kabir
D. M. A. Hussain	Kamlesh Dutta
Jason Levy	Sobha Lalitha Devi
Sri Chusri Haryanti	Mamoon Rashid

Gaurav Sharma
Sandip Goyal
Rohit Handa
Bindeshwar Singh
Amanpreet Kaur

Sonia Sharma
Rajesh Chauhan
Balvir Singh Thakur
Gurpreet Singh
Rohit Bajaj

Contents – Part I

Machine Learning and Deep Learning

A Novel Smart Facial Features for Real-Time Motorists Sleepiness Prediction and Alerting System Using Hybrid Deep Convolutional Neural Network in Computer Vision 3
   G. A. Senthil, R. Prabha, S. Sridevi, R. Deepa, and S. Shimona

Enhancing Power Transformer Reliability with Machine Learning-Based Fault Detection and Data Analysis 17
   Abha Sharma, Vaasu Bisht, Aarav Rajput, Vansh Dugar, and Anand Lahoti

Type 2 Diabetes Prediction Using Machine Learning: A Game-Changer for Healthcare 34
   Sadhana Singh, Priyanka Sharma, and Pragya Pandey

Classification of Brain Tumors in MRI Images Using Deep Learning 45
   Rahul Namdeo Jadhav and G. Sudhagar

Improving Breast Cancer Prediction with Validation: Optimized Feature Extraction with Spider Monkey Optimization and Validation with Cutting-Edge Classifiers 57
   Emmy Bhatti and Prabhpreet Kaur

Diabetes Prediction Precision: Evaluating XGBoost and LightGBM Performance with IQR Preprocessing 74
   K. Kotaiah Chowdary and K. N. Madhavi Latha

Enhancing Cardiovascular Health Prediction: A Machine Learning Perspective 87
   Ratnam Dodda, Abhishek Reddy Bonam, Srinidhi Sakinala, and Prithika Reddy Dendhi

Measuring AI's Impact on HR Strategies and Operational Effectiveness in the Era of Industry 4.0 97
   K. Devi, Jagendra Singh, Neha Garg, Mohit Tiwari, Nazeer Shaik, and Muniyandy Elangovan

Implementation of Deep Learning and Machine Learning for Designing and Analyzing IDS (Intrusion Detection System) Through Novel Framework 108
   Kiranjeet Kaur and Jaspreet Singh Batth

Performance Evaluation of Medicinal Leaf Classification Using DeepLabv3 and ML Classifiers 124
   Ashwin Kumar Bodla and Rama Krishna Damodara

Android App Permission Detector Based on Machine Learning Models 135
   Jaikishan Mohanty and Divyashikha Sethia

Histopathological Analysis Advancements in Deep Learning for the Diagnosis of Lung and Colon Cancer with Explanatory Power via Visual Saliency 148
   Seema Kashyap, Arvind Kumar Shukla, and Iram Naim

Deep Learning Approaches for Chest X-Ray Based Covid-19 Identification 159
   R. Prabu and P. Dhinakar

Pattern and Speech Recognition

Bi-LSTM Based Speech Emotion Recognition 173
   Pagidirayi Anil Kumar and B. Anuradha

Real Time Voice Language Interaction 192
   Gagan S. Yadav, J. Vimala Devi, Tanmai Jain, H. D. Harshith, and B. N. Neha

Exploring In-Context Learning: A Deep Dive into Model Size, Templates, and Few-Shot Learning for Text Classification 207
   Areeg Fahad Rasheed, Safa F. Abbas, and M. Zarkoosh

Unlocking Complexity: Traversing Varied Hurdles in Punjabi Newspaper Recognition System 216
   Atul Kumar and Gurpreet Singh Lehal

Internet of Things (IoT)

Patient Biomedical Monitoring - Remote Care System 233
   D. Sri Chandana Charudatta, B. Sidharth Reddy, CH. Pavan Sai, and S. Srinivas

An Energy-Efficient Clustering Approach for Two-Layer IoT Architecture 245
   Annu Malik and Rashmi Kushwah

Blockchain Research in Healthcare: A Bibliometric Review 257
   S. Neeraj and Anupam Bhatia

Enhancing IoT Security Through Hierarchical Message-Passing Graph
Neural Networks: A Trust-Driven Strategy for Identifying Malicious Nodes 273
   C. Senthil Kumar and R. Vijay Anand

Author Index 287

Contents – Part II

Data Science and Data Analytics

Generalized Material Flow Model for Hot Deformation Processing Based on Power Law 3
  Krishna Kant Pathak, Abha Sharma, and Denny Ben

Smallest Path Planning for Robots Using Simple ACO and Ant System Algorithms 16
  Gurpreet Singh, Amanpreet Kaur, Shilpa, and Aashdeep Singh

An Asymmetric Single Channel Encryption Mechanism Using Unequal Modulus and QZS Algorithm in Fractional Fourier Transform 25
  Pankaj Rakheja and Neeti Kashyap

Evaluation Criteria for Test Suites in Software Developed Through Test-Driven Development 39
  Tara Rani Agarwal and Nikita Thakur

LCB-eGyan Model: A Dynamic Cloud Architecture for e-Content Recommendation 53
  Nidhi Goyal

Gender and Ethnicity Recognition System Based on Convolutional Neural Networks 66
  S. Harishwaran, Rakoth Kandan Sambandam, and R. Gokulapriya

Radiographic Evaluation for KNEEXNET Using Ensemble Models 77
  P. Pal, R. Dhara, A. Bal, and R. Bhattacharya

Comparative Study of Quantum Computing Tools and Frameworks 87
  Muhammad Hamid, Bashir Alam, and Om Pal

Security Evaluation and Oversight in Stock Trading Using Artificial Intelligence 105
  Devadutta Indoria, Jagendra Singh, Neha Garg, Mohit Tiwari, B. N. Karthik, and Nazeer Shaik

A Dual Approach with Grad-CAM and Layer-Wise Relevance Propagation for CNN Models Explainability 116
  Abhilash Mishra and Manisha Malhotra

CLAM – CNN and LSTM with Attention Mechanism for DNS Data Exfiltration Detection 130
    Jisha Joy and Shivani Jaswal

Comparison of Various Data Mining Techniques for Fake Profile Detection on Twitter 145
    Swati Gupta and Sonal Saurabh

Strategies for Gen AI Enterprise Adoption 154
    Tirupathi Rao Dockara and Pradeep Rajagopal Kirthivasan

Role of Deep Learning in Skin Cancer Diagnosis: A Deeper Insight 176
    Roohi Singh and Rakesh Chandra Gangwar

Communication, Network and Security

Smart Education-Filtering Real Time Stream 191
    Shreyans Jain, Sagar Soni, Shubbham Jain, and Deepali Kamthania

RNA and Audio-Based Workflows Scheduling by Multi-Objective Optimization in Cloud Environment 204
    Vivek Kumar and Ram Krishan

Group Authentication and Key Management Protocol with Chinese Remainder Theorem in Multicast Communication 217
    Konjengbam Roshan Singh and Divyashikha Sethia

Network Traffic Analysis for Intrusion with Zero Trust 230
    Nitin Ravi Gautam, Harsh Mishra, Arun Kumar Singh,
    and Sandeep Saxena

Enhanced Image Deblurring with the Fusion of Generative Adversarial Networks and Transformer Models 246
    Arti Ranjan and M. Ravinder

Analysis of Various Routing Protocols for Air Pollution Monitoring Systems in Wireless Sensor Networks 260
    Arzoo and Kiranbir Kaur

Intelligent Toll Management Using Limited CNN-Based Vehicle Classification 271
    Jagdish Chandra Patni, Sudhir Kumar Rajput,
    and Nilesh Bhaskarrao Bahadure

Optimizing Software Demands Using Fuzzy-Based Evaluation Techniques 282
  Rubi, Jagendra Singh, Dinesh Prasad Sahu, Mohit Tiwari,
  Nazeer Shaik, and A. K. Shrivastav

Enhancing Workflow Efficiency: Innovative Workload Clustering in Cloud
Environments 293
  Monika Yadav and Atul Mishra

Author Index 309

Machine Learning and Deep Learning

A Novel Smart Facial Features for Real-Time Motorists Sleepiness Prediction and Alerting System Using Hybrid Deep Convolutional Neural Network in Computer Vision

G. A. Senthil[1](✉), R. Prabha[2], S. Sridevi[3], R. Deepa[3], and S. Shimona[4]

[1] Department of Information Technology, Agni College of Technology, Chennai, India
senthilga@gmail.com
[2] Department of Electronics and Communication Engineering, Sri Sai Ram Institute of Technology, Chennai, India
[3] Department of Computer Science and Engineering, Vels Institute of Science, Technology and Advanced Studies, Chennai, India
sridevis.se@velsuniv.ac.in
[4] Department of Computer Science and Engineering, Agni College of Technology, Chennai, India

Abstract. Motorists Sleepiness prediction is a process of detecting when an Operator is experiencing Sleepiness or fatigue while driving a vehicle. This is an important safety feature, as Sleepy driving can lead to accidents and injuries. There are several methods used to predict Operator Sleepiness, including physiological monitoring, behavioural monitoring, and hybrid methods. Physiological monitoring methods like CNN involve measuring the Operator's physiological signals, such as image or video frame processing from a camera. These frames can provide information about the Operator's level of alertness and can be used to detect Sleepiness. Behavioural monitoring methods that are DCNN on the other hand, involve observing the Operator's behaviour, such as comparing the frames with the processed dataset. This information is mainly used to detect Sleepiness. Hybrid methods combine physiological and behavioural monitoring methods of CNN and DCNN and added to the fuzzy logic algorithm makes an HDCNN (Hybrid Deep Convolutional Neural Network) to provide a more comprehensive assessment of the Operator's level of Sleepiness and improves the accuracy. This article explains techniques to spot the lips and eyes in a video taken during a research project by the Indian Institute of Road Safety (IIROS). A footage of the transition from awake to fatigued to drowsy will be captured using the digital camera. The Proposed algorithm's function is to locate the face recognized in a captured video image. The face region is used mostly for its ability to act independently.

Keywords: Convolutional Neural Network (CNN) · Hybrid Deep Convolutional Neural Network (HDCNN) · Fuzzy Logic Algorithm · Eye Aspect Ratio (EAR) · Computer Vision · Haar wavelet

© The Author(s), under exclusive license to Springer Nature Switzerland AG 2025
M. Malhotra (Ed.): IETCIT 2024, CCIS 2125, pp. 3–16, 2025.
https://doi.org/10.1007/978-3-031-80839-5_1

1 Introduction

Every individual require sleep, and lack of sleeping leads to lethargy, sluggish reflexes, blurred focus, and deviation, all of these factors limit a human's ability to make the kinds of decisions needed when driving a type of vehicle [1]. The World Health Organization (WHO) estimates that accidents cause over 1.5 million injuries or fatalities annually. Some of them drive excessively quickly, cut through red lights, and cross highways in violation of the traffic regulations [2]. Their brakes and tyres are mechanically flawed as well. The novel objective of the research is to provide a comprehensive sleepiness monitoring system as a means of mitigating these issues in order to reduce fatal incidents. This model's accuracy is 95%. This strategy makes advantage of deep learning as well [4]. The technique incorporates both hybrid deep learning algorithm and machine learning based on augmented reality with computer vision, which are two disciplines of artificial intelligence that enable. The user can train the algorithm to forecast output within a defined range. This kind of technology bridges the gap between people and robots [5].

Computer vision is a technology designed to gather and interpreting images for image preprocessing. It facilitates the collection and preprocessing of trained data to provide information, and it also makes use of other system-critical technologies. Because of their intuitive graphical user interface (GUI), operating systems are easy for users to use. Where you may easily construct your own panels to produce the output of a certain model. We're using two distinct models here for safety reasons: drowsiness monitoring and driver identification of face characteristics. The model's output is to warn and prediction of the driver falls asleep, switches lanes, or the automobile is not properly technology maintained. The goal of this research is to protect drivers by incorporating this technology into automobiles [6].

Human beings have constantly developed tools and ways to make lifestyle simpler and safer, the fact that either tedious tasks like travelling to work or more thrilling ones like flying. Alongside the increase in generation, modes of transportation established prominence, and our reliance on them rose substantially. As a result, our lives as we know them have changed dramatically. We can go to places at a rate that our forebears could not have imagined conceivable. We are able to travel to locations at a pace that not even our grandparents would have thought possible. Today, practically everyone on our planet uses a form of transportation every day. Few individuals are rich enough to own vehicles, and others rely on public transportation. Nevertheless, there are a few rules and standards of conduct for those who exert pressure, regardless of their standing in society. One of them is maintaining consciousness, alertness, and energy while driving [7].

By ignoring our obligations in the direction of safer travel, we have allowed vast volumes of harm to merge with such brilliant ingenuity each year. For most seniors, it can seem like a minor detail, yet adhering to the law and acting responsibly on the street have major ramifications. Even while a motor vehicle uses the most of its electrical power on pavement, in careless hands, it may be devastating and occasionally, such negligence puts lives of individuals in danger. One example of inexperience is failing to understand that we are too tired to drive. Several academics have published publications on driver tiredness detection devices in attempt to identify and prevent an unfavourable consequence from such conditions. However, several considerations and tests performed with the gadget are occasionally inadequately precise. This project was conducted in order to better their implementations and similarly optimise the solution by offering analytics and any other perspectives that is relevant to the situation at hand. [8].

1.1 Problem Definition

Sleep deprivation is an issue of security that the world has not lately really addressed. Sleepiness, unlike alcohol or drugs, may be difficult to measure or evaluate due to the way it functions. While weariness or drowsiness is not quantifiable or identified and is a common problem, drugs and alcohol have prediction clear key indications and tested that can be acquired without vehicle problems, which makes them clearly detectable and avoidable. The most probable remedies to this problem are to concentrate on events that are caused by driver drowsiness and to encourage drivers to admit it when necessary. The final outcome cannot be accomplished without the former, because riding for long periods of time may be quite useful. As a result, tiredness monitoring techniques are required for the prevent accident of automobile vehicle and their drivers. The first option is a lot more challenging and costly to do.

Fig. 1. Fatigued Driver

Much of the background work on facial recognition focuses on identifying discrete features such as the appearance of eye, mouth, and head, and defining the position, shape, and relationship of these features. Face popularity has emerged as a crucial problem in lots of programs, together with security structures, credit score card verification, and criminal identity. The capability to version a particular face and the potential to distinguish it from a big variety of stored face fashions can also beautify protection and stumble on faces, in comparison, can be important to discover. Facial recognition in pix is very useful for automating shade film development, because the effect of a couple of enhancement and noise discount methods depends on the photograph content material [9]. Figure 1 shows Fatigued Driver facial recognition.

Haar cascade xml is a method primarily based on machine getting to know (ml) in which lots of tremendous and negative images are used to teach the classifier. Superb pictures are the ones pictures that incorporate the photographs which we need our classifier to become aware of as an example: face and negative pix are the images of the whole thing else i.e., they are the pictures which do now not contain the item we need to discover. This helps us in face detection so that the device can efficaciously come across a face so that it will method the face reputation method. The Haar Cascade classification is based on Harvowlet technology to analyse the pixels in an image in sections by function. It uses the concept of "comprehensive image" to calculate "features "Need to know. Haar cascade makes use of the ada-increase mastering set of rules to pick a small range from a massive set of statistics to provide a powerful result of essential capabilities from a big set and cascading method is used to identify the face within the picture [10].

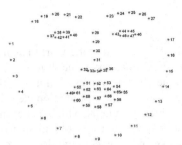

Fig. 2. D-lib's 68 Points Face

D-lib is a toolkit used for making Realtime machine learning applications and information evaluation packages. It's used for face detection/ face recognition and facial landmark detection. The prefrontal facial detection in d-lib performs admirably since it is quite straightforward and works right out of the box. Figure 2 shows D-lib has a model with 68 points. Here, we can see that it multiplies from 1 to 68. Yet, there are times when we only need a few of the 68 feature points. As a result, we may tailor those factors to fit our needs. For instance, in our machine, we just employ criteria 38 to 58 to calculate the EAR. Basically, kind of two ways to accidentally capture landmarks.

Recognition of Faces: Recognition of faces is an initial technique that the predict a human face and returns a rectangle-shaped points in the coordinates x, y, w, and h.

Facial Reference: We must first find a face's placement in the image before moving on to the locations within the shape of a rectangle.

2 Related Work

Zheng, g et. al (2008), In order to reduce the act threat problem, this paper suggests a new generation of modelling driving force fatigue supported by statistics fusion approach with multiple eyelid motion characteristics—particular minimal squares regression. Where there is a strong correlation between the eyelid movement features and the tendency to drowsiness [1]. The tentative correctness and robustness of the correspondingly established version have been confirmed, indicating that it offers an alternative method of concurrent multi-fusing to increase our capacity to recognize and anticipate sleepiness.

Friedrichs, f et. al (2010), In this research, it is suggested that the motivation eye measurements be started to detect sleepiness in a computerized laboratory or experiment setting. Modern optical productivity fatigue tests are categorized as supported assessment measures. These measurements are backed by statistics and a categorization approach for a 90-h huge dataset of drives on a crucial road [2]. The results demonstrate that for certain drivers, eye-tracking results in detecting tiredness works longer. Blink detection still has issues when used by people wearing sports glasses and poor lighting situations, despite some of the suggested enhancements. In summary, digital camera-based sleep measures offer valuable assistance for sleepiness, but relying solely on advice is unreliable.

Armingol, j. M et.al (2011), In order to reduce the number of deaths, an application for a sophisticated driving support system that automatically identifies driver tiredness and also makes driving distraction easier is presented in this study. The sleepiness index is calculated using machine learning algorithms that use visual data to recognize determine, and study each driver's face characteristics and eyes [3]. Due to the near-infrared illumination system, this real-time gadget can function at night. The proposed set of principles is ultimately verified by instances of various motivation force photos taken overnight in a real car.

Zhang, W., Cheng, B et.al (2013), explore the address concerns caused by changes in illumination and driving position, this article introduced a robust set of eye detection criteria to a non-profit sleepiness detection approach that makes use of eye tracking and image processing [4]. The six measurements proportion of yield closure, maximum final period frequency of eyelids, average eye level releasing frequency, eye velocity starting, and eye pace are computed, which provides 86% accuracy.

3 Proposed Work and Methodology

3.1 Convolutional Neural Network (CNN)

A vehicle's sleepiness detection system may be successfully implemented using a CNN (Convolutional Neural Network). To identify indicators of driver intoxication and warn them to avoid collisions, the system can combine computer vision methods with machine learning algorithms. Here's a high-level overview of the concept and algorithm for a driver drowsiness system using CNN:

Step 1: Data Collection: The initial phase entails gathering an enormous collection of images and videos of automobile drivers in various states, including alert and sleepy states. These images or videos should be labelled to indicate the corresponding state.

Step 2: Preprocessing: The collected images or video frames need to be pre-processed to enhance the relevant features. This may involve resizing, normalization, and filtering techniques to improve the quality and consistency of the data.

Step 3: Feature Extraction: The CNN is responsible for automatically learning relevant features from the pre-processed images. This is achieved through a series of convolutional, pooling, and activation layers. The convolutional layers capture local patterns and features, while the pooling layers down sample the spatial dimensions and reduce computational complexity.

Step 4: Training: Sets for validation and training purposes are created from the labelled dataset. The CNN is trained using the training set, and the validation set is used to monitor the model's performance and prevent overfitting. The network learns to recognize patterns and features associated with drowsiness during this phase. The training process involves optimizing the network's parameters using gradient descent and backpropagation.

Step 5: Testing and Evaluation: Once the CNN is trained, it can be tested on a separate dataset that was not used during training. This dataset should contain real-time images or video frames. The CNN processes these frames and predicts the driver's drowsiness state. The predictions can then be evaluated against the ground truth labels to measure the accuracy and effectiveness of the model.

Step 6: Alert System: Based on the predictions made by the CNN, an alert system can be triggered to notify the driver if drowsiness is detected. This can be achieved through various mechanisms such as visual alerts, audible alarms, or vibrations.

Step 7: Continuous Monitoring: The system should continuously monitor the driver's behavior and update the drowsiness predictions in real-time. This allows for timely alerts and ensures the system adapts to changes in the driver's state.

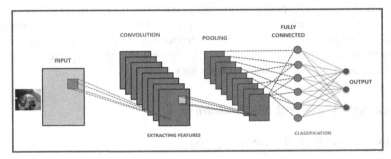

Fig. 3. Architecture of CNN (Convolutional Neural Network)

It's important to note that the above steps provide a general outline, and the actual implementation details may vary depending on the specific requirements and constraints of the system. Additionally, the CNN architecture and hyperparameters, as well as the alert mechanisms, can be further optimized through experimentation and fine-tuning. Figure 3 shows architecture of CNN (Convolutional Neural Network) Model.

Figure 4 shows Deep Convolutional Neural Network is referred to as DCNN, the artificial neural network type is frequently employed for computer vision tasks. If you want to apply the DCNN algorithm to detect driver drowsiness, you can follow these general steps.

Step 1: Dataset Collection: Gather a dataset of images or videos of drivers exhibiting both drowsy and alert states. It's important to have a diverse and representative dataset to train a robust model.

Step 2: Data Preprocessing: Resize the photos, normalize the pixel values, and do any necessary adjustments prior to processing the dataset. Additionally, label the images or videos to indicate whether the driver is drowsy or alert.

Step 3: Model Architecture: Create a DCNN architecture that is appropriate for the objective of detecting driver intoxication. Convolutional layers, pooling layers, and fully linked layers are frequently stacked in this manner. You may also consider using pre-trained models such as AlexNet, VGG, ResNet50, or Inception, and fine-tune them for your specific task.

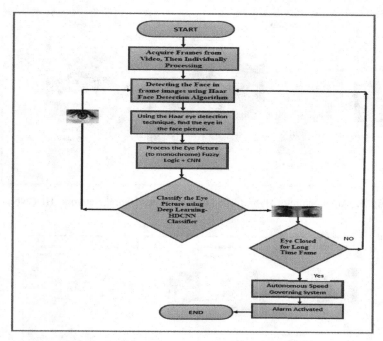

Fig. 4. Function Block Diagram of Deep Convolutional Neural Network (DCNN)

Step 4: Training: Generate training and validation sets from the dataset. The data used for training should be fed into the DCNN model, and the model parameters should be optimized using the right optimization technique (for example, stochastic gradient descent) and loss function (for example, binary cross-entropy). Training the model across several iterations while keeping an eye on results of validation to prevent excessive fitting.

Step 5: Evaluation and Testing: After the model has been trained, assess its effectiveness using a different test set that wasn't utilized during training. Determine criteria including accuracy, precision, recall, and F1 score to evaluate the model's performance at spotting sleepiness.

Step 6: Deployment: Once the model performs well on the test set, you can deploy it for real-time drowsiness detection. This may involve integrating it into an application or system that can continuously analyse video streams or images from a camera in a vehicle.

Understand that other elements like face identification, facial landmark detection, and surveillance algorithms may be required to construct a full DCNN-based sleepiness prediction technology. These components can help isolate and analyse the driver's facial region for more accurate drowsiness detection. It's crucial to remember that while HDC-NNs can be successful in detecting sleepiness, the algorithm's effectiveness is dependent on the dataset's excellence, the model's design, and the training procedure. Continuous monitoring and improvement are necessary for robust and reliable drowsiness detection as shown in Fig. 5 above and Fig. 6 shows max polling [11].

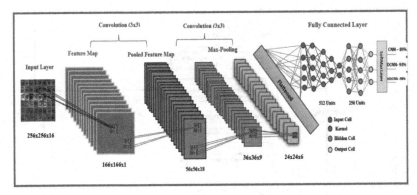

Fig. 5. Architecture of Hybrid Deep Convolutional Neural Network (HDCNN)

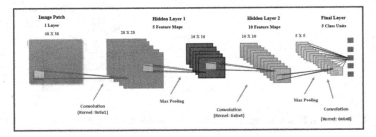

Fig. 6. Max Pooling Architecture

In the Sleep deprivation Recognition System, the initial module. This device analyses the driver's eyes and facial features to determine the degree to which they are sleepy. The process begins with recognition of facial features in order to identify an eye blink. Once the system has identified the driver's eyes, it determines the velocity of the eye blink and outputs the appropriate data.

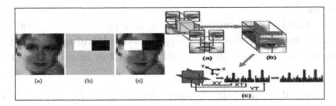

Fig. 7. Local Binary Pattern and Training Haar Cascades

Figure 7 shows the OpenCV is used by the HAAR cascade approach to identify faces of people. The nose is brilliant, but the human eye is black. So, utilizing OpenCV, the Haar Cascade approach is employed to retrieve the facial features. With this technique, the eye's location on the human body's dimension is recognized. The face is continually tracked by the system until the user turns it off. When the data is supplied on a plane, the mean shift method is utilized to locate the center of the picture for image distributions [12]. A Local Binary Patterns (LBP) structuring method may determine the identity of each pixel in an image by the thresholding the actual size of each pixel and seeing the result as a binary value. The LBP texture operator is a well-liked technique in many applications because to its strong discriminative ability and computational simplicity. It may be viewed as a reconciling method for texture analysis's statistical and structural models, which are frequently at variance with one another. The LBP operator's capacity to adjust to recurring changes in grayscale [9].

I need to determine the picture's location and radius in order to attain the dense portion of the image. D-lib is an open-source library that is used for both eye recognition and eye aspect calculation. The eye's feature was proportion. Ratio has a threshold of 0.3; if this threshold remains constant, the system is going to presume that the eyes are open, and if it decreases below this threshold, the computer will assume that the subject is in a sleepy condition [13–15].

3.2 Eye Aspect Ratio (EAR)

Figure 9 shows EAR, is the ratio of the length and the width of the eyes. The length and width of the eyes is calculated by means of averaging over the two horizontal and vertical lines throughout the eyes as illustrated [16, 17].

Our speculation turned into that once a man or woman is drowsy, their eyes are probable to get smaller and they are probable to blink extra. Based totally on this speculation, we anticipated our model to expect the class as drowsy if the EAR for a character over successive frames declined i.e. Their eyes began to be close or they were blinking faster.

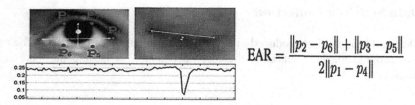

$$\text{EAR} = \frac{\|p_2 - p_6\| + \|p_3 - p_5\|}{2\|p_1 - p_4\|}$$

Fig. 8. Eye Landmarks and EAR

The above Fig. 8 shows the identification system faces its subsequent module. Here, we're use object detection algorithms to identify facial traits. We capture both favourable and unfavourable photographs, then feed the algorithm with the information.

Quick recognizing an object using a boosted cascade of simple functions is a set of rules for object recognition that is used in machine learning and was proposed by Paul viola and Michael Jones in the article "Fast object detection the usage of a boosted cascade of easy functions" in 2002. The data set being provided is now divided into two phases: training and testing for the model. We'll utilize 80% of the data for training and 20% for testing. We are developing a GUI (Graphical user interface) utilizing python Tkinter GUI toolkit to communicate with the system while maintaining the user experience modest. Figure 9 shows prediction of eye in computer vision.

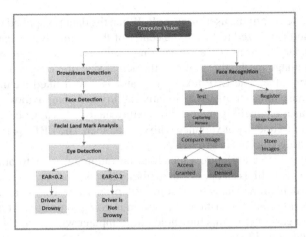

Fig. 9. System Architecture of Prediction of Eye in Computer Vision

4 Feature Extraction

Just briefly said earlier, were set out to develop the necessary skills for our class model based on the facial landmarks that we retrieved from the frames of the films. While testing several hypotheses and functions, we came to the following conclusions about our final models: eye element ratio, mouth thing ratio, learner circularity, and eventually, mouth element ratio above eye thing ratio.

5 Data Set/Data Collection

The Dataset is gathered or produced in accordance with our research model. The more data we provide machine learning, the more dependable it is, and the smarter the algorithm gets.

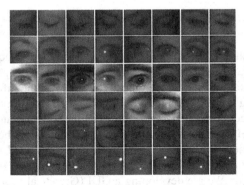

Fig. 10. Dataset Classification of Drowsing Eyes

Any classification algorithm may be used to train the model, and if there is enough data, the system will learn more quickly and with more accuracy. We have special record settings on this paper to instruct the version. The first model is a system for detecting sleepiness, and it uses data photos from Kaggle. A platform that is open-source that offers datasets is called Kaggle. Photographs with the eyes wide and narrow make up the majority of the collection of photographs for this model. The facial recognition technology in the second model recognizes a person's visage for security reasons. The driver's face photos that were photographed are included in this collection. The hybrid algorithm may also be trained using live images, improving the accuracy of the system. In this model, pictures of various drivers are taken and saved in folders for various categories that are designated as the model's dataset. Figure 10 shows the dataset classification of drowsing eyes. The simulation has implementation using python.

6 Results and Discussion

In this system we're adding up two different models, that is drowsiness detection and face recognition for safety and security purposes of the driver. First, we will be popping up with the GUI window which has two buttons, one is drowsiness detection and other is face recognition system.

If we click the button drowsiness detection it will redirect you to the alert system. Here GUI consists of the live feed which is continuously fed to the camera with the bounding boxes detects the eyes and the face of the operator and if the operator closes his eyes for 3–4 s an alarm system will be activated.

Fig. 11. Output of Drowsiness Prediction

If you will be selecting face recognition you will be redirected to the face recognition model. Once camera captures your image it will verify with the data provided and it recognizes the data and displays the driver's name. Figure 11 shows simulation of the computer vision image capture face recognitions for drowsiness prediction.

Table 1 shows above performance of the proposed simulation with various parameter of different deep learning algorithm in the process of prediction of facial drowsiness of

Table 1. Prediction Accuracy of Facial Features Using Algorithm

Deep Learning Classifiers	Accuracy
CNN	89%
DCNN	92%
HDCNN	98%

motorist with accuracy and efficiency values in the percentage of CNN is 89%, DCNN is 92% and HDCNN is 98%. The overall best accuracy has predicted HDCNN algorithm is consider an efficient methodology.

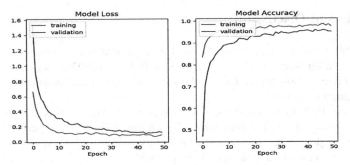

Fig. 12. Training Validation of Model Loss and Accuracy for HDCNN

The deep learning model's model accuracy and model loss are displayed in the graph above Fig. 12. It demonstrates that when epochs are increased, model accuracy grows. In a similar vein, model loss decreases as epoch increases.

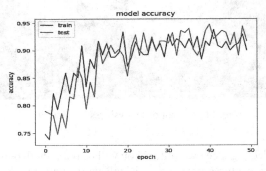

Fig. 13. Training and Testing of Model Accuracy epoch for HDCNN

Figure 13 shows the efficiency and accuracy resultant motorist eye tiredness prediction algorithm creation phase has been implemented using Matplotlib using python.

CNN is used to build an algorithm that classifies types depending on datasets. The framework consists of five layers, all three of which are hidden, and uses the SoftMax activation function and the Adam optimizer to accomplish HDCNN classification. The model is trained over 50 iterations, and after achieving an accuracy rate of 98%, the loss can be quickly reduced. The simulation's precision is insufficient for real-time application because it is used for ocular imaging. Virtual Reality (VR) and Augmented Reality (AR) Technology is integrated in surgical operations using mixed reality in medicine. It improves the surgeon's perception and gives crucial information during the procedure by fusing computer-generated images and data with real-time imaging of the surgical field. Here are a few techniques frequently employed in mixed reality-based medical surgery.

7 Conclusion and Future Work

The motorist's security features have been evaluated. First, we developed a method for alerting the driver if he or she experiences tiredness for more than 3 to 4 s. This allows the driver to choose whether to continue driving or taking a rest. As opposed to the first model, which uses facial recognition to identify the motorist and provide entry, the second model secures our car. The facial recognition system is extremely beneficial for preserving the security of the car avoiding vehicle thefts, and the drowsiness prediction system may be installed in every automobile so that we might avoid vehicle accidents and minimize the mortality of ratio which is caused by tiredness.

Compared to innovative algorithms, HDCNN delivers better and even more exceptional outcomes when it comes to solving the difficulties using Computer Vision Technology Design. By demanding a high efficiency of objective function evaluations, HDCNN is able to arrive to the accuracy prediction functions that are the best overall, demonstrating its computational efficiency. In the chosen mathematical test functions, HDCNN outperforms the comparison algorithms because it converges to the preset tolerance of the world's best in a quicker and more effective manner. The proposed work hybrid deep learning model smart facial feature for driver drowsiness prediction analysis for safety measure implemented using deep learning algorithm such as Convolution Neural Network (CNN), Deep Convolution Neural Network (DCNN) and Hybrid Deep Convolution Neural Network (HDCNN) accuracies are 98%.

Further, using a compatible mouth dimension threshold evaluation, we can identify the driver's jaw area, calculate the mouth aspect ratio (MAR), and then utilize it to recognize frequent yawning and notify the driver. This would broaden the scope of the study and improve our ability to recognize weariness. We may potentially improve the sensors and employ night vision sensors to better discern scenes in low light. We upgraded with AR/VR.

References

1. Su, H., Zheng, G.: A partial least squares regression-based fusion model for predicting the trend in drowsiness. IEEE Trans. Syst. Man Cybern. Part A Syst. Hum.Cybern. Part A Syst. Hum. **38**(5), 1085–1092 (2008)

2. Friedrichs, F., Yang, B.: Camera-based drowsiness reference for driver state classification under real driving conditions. In: Proceedings of the 2010 IEEE Intelligent Vehicles Symposium, pp. 101–106. IEEE (2010)
3. Flores, M.J., Armingol, J.M., de la Escalera, A.: Driver drowsiness detection system under infrared illumination for an intelligent vehicle. IET Intel. Transport Syst. 5(4), 241–251 (2011)
4. Zhang, W., Cheng, B., Lin, Y.: Driver drowsiness recognition based on computer vision technology. Tsinghua Sci. Technol. 17(3), 354–362 (2013)
5. Mbouna, R.O., Kong, S.G., Chun, M.G.: Visual analysis of eye state and head pose for driver alertness monitoring. IEEE Trans. Intell. Transp. Syst.Intell. Transp. Syst. 14(3), 1462–1469 (2013)
6. Suganthi, P., Boopathi, D.: Analysis of cognitive emotional and behavioral aspects of Alzheimer's disease using hybrid CNN model. In: International Conference on Computer, Power and Communications (ICCPC), Chennai, India, pp. 408–412 (2022). https://doi.org/10.1109/ICCPC55978.2022.10072126
7. Senthil, G.A., Prabha, R., Priya, R.M.: Classification of credit card transactions using machine learning. In: International Conference on Computer, Power and Communications (ICCPC), Chennai, India, pp. 219–223 (2022), https://doi.org/10.1109/ICCPC55978.2022.10072269
8. Nisha, S.A., Snega, S., Keerthana, L.: Comparison of machine learning algorithms for hotel booking cancellation in automated method. In: International Conference on Computer, Power and Communications (ICCPC), Chennai, India, pp. 413–418 (2022). https://doi.org/10.1109/ICCPC55978.2022.10072135
9. Sridevi, S., Monica, K.M.: Third generation security system for face detection in ATM machine using computer vision. In: International Conference on Computer, Power and Communications (ICCPC), Chennai, India, pp. 143–148 (2022). https://doi.org/10.1109/ICCPC55978.2022.10072096
10. Tadesse, E., Sheng, W., Liu, M.: Driver drowsiness detection through HMM based dynamic modeling. In: Proceedings of the 2014 IEEE International Conference on Robotics and Automation (ICRA), pp. 4003–4008. IEEE (2014)
11. Babaeian, M., Bhardwaj, N., Esquivel, B., Mozumdar, M.: Real time driver drowsiness detection using a logistic-regression-based machine learning algorithm. In: Proceedings of the 2016 IEEE Green Energy and Systems Conference (IGSEC), pp. 1–6. IEEE (2016)
12. Yan, J.J., Kuo, H.H., Lin, Y.F., Liao, T.L.: Real-time driver drowsiness detection system based on PERCLOS and grayscale image processing. In: Proceedings of the 2016 International Symposium on Computer, Consumer and Control (IS3C), pp. 243–246. IEEE (2016)
13. Zhao, X., Wei, C.: A real-time face recognition system based on the improved LBPH algorithm. In: Proceedings of the 2017 IEEE 2nd International Conference on Signal and Image Processing (ICSIP), pp. 72–76. IEEE (2017)
14. Reddy, B., Kim, Y.H., Yun, S., Seo, C., Jang, J.: Real-time driver drowsiness detection for embedded system using model compression of deep neural networks. In: Proceedings of the IEEE Conference on Computer Vision and Pattern Recognition Workshops, pp. 121–128 (2017)
15. Nithyashri, J., et al.: A novel analysis and detection of autism spectrum disorder in artificial intelligence using hybrid machine learning. In: International Conference on Innovative Data Communication Technologies and Application (ICIDCA), Uttarakhand, India, pp. 291–296 (2023). https://doi.org/10.1109/ICIDCA56705.2023.10099683
16. Qu, X., Wei, T., Peng, C., Du, P.: A fast face recognition system based on deep learning. In: Proceedings of the 2018 11th International Symposium on Computational Intelligence and Design (ISCID), vol. 1, pp. 289–292. IEEE, December 2018
17. Mehta, S., Mishra, P., Bhatt, A.J., Agarwal, P.: AD3S: advanced driver drowsiness detection system using machine learning. In: Proceedings of the 2019 Fifth International Conference on Image Information Processing (ICIIP), pp. 108–113. IEEE, November 2019

Enhancing Power Transformer Reliability with Machine Learning-Based Fault Detection and Data Analysis

Abha Sharma[✉], Vaasu Bisht, Aarav Rajput, Vansh Dugar, and Anand Lahoti

School of Computing Science and Engineering, VIT Bhopal University, Bhopal-Indore Highway, KothriKalan, Sehore, Madhya Pradesh 466114, India
{abhasharma,vaasu.bisht2021,aarav.rajput2021,vansh.dugar2021, anand.lahoti2021}@vitbhopal.ac.in

Abstract. For stepping up and down voltage levels, in electrical power infrastructure power transformers play an important role in efficient transmission and distribution of electricity However, transformer failures are a common problem that can result in power outages, equipment damage, and safety risks. The study involved extensive experimentation with various preprocessing techniques, feature selection methods, resampling techniques, and model evaluation metrics to identifying when transformers are likely to fail, this can help prevent transformer failures incidents and minimize their impact. This study focused on developing an effective predictive model to identify transformer failures using the Winding Temperature Indicator (WTI) as the target feature.

Keywords: Machine learning · feature selection · power transformers · winding temperature indicator · resampling

1 Introduction

In the context of enhancing power transformer reliability, artificial intelligence mainly machine learning techniques can play a critical role in improving the accuracy and effectiveness of fault identification. Traditional methods of fault detection in power transformers rely on the expertise of engineers and their ability to interpret monitoring data and detect patterns indicative of faults. This approach can be time-consuming and may not always be accurate [1]. Machine learning-based fault detection can automate the process of fault diagnosis by extracting knowledge from large datasets and identifying hidden relationships between monitoring data and the health states of transformers. It is possible to detect patterns and anomalies that may indicate the presence of faults or potential issues with the transformer using machine learning [2].

2 Research Elaborations

The following section outlines the methodology employed in developing a supervised machine learning model for predicting transformer failure based on the Winding Temperature Indicator (WTI). This study aims to address the critical issue of monitoring temperature levels in transformer windings and detecting instances where the WTI threshold is exceeded, leading to potential transformer failure. By utilizing machine learning various models and pre-processing technique, we made a binary classification model that can forecast transform failures.

2.1 The Data

In this study, a total of 20,465 records of IoT sensor data were collected to gather information on parameters such as Phase Line readings, Current Line, Voltage Line readings etc. from June 25th, 2019 to April 14th, 2022. The data was collected at 15-min intervals, resulting in a large dataset that provides a comprehensive view of the parameters being studied. The data is collected using two different IoT sensor device types which collect data in various formats from varying data sources. Because of improper integration and technical limitations involving the system design between the two types of sensors, the collected data is stored in two separate sheets and used for analysis. The first dataset contains data related to different indicators, while the second dataset contains data related to current and voltage. Synchronization of both sensor devices will be a critical aspect of the data collection. Each measurement will be accurately timestamped to maintain data integrity and enable proper analysis. By ensuring synchronization, the collected data from multiple sensors can be effectively correlated for comprehensive analysis.

When it comes to analysing and processing the collected data, it's necessary to combine both datasets. One way to achieve this is by using the datetime column, which is a common column that exists in both files. By concatenating the datasets based on the datetime column, we can ensure that the data is correctly aligned, and that no information is lost. This process is particularly important when performing machine learning on the collected data. To make accurate predictions majority of machine learning algorithms needs large amount of data.

2.2 The Data Preprocessing

Pre-processing methods are used to make sure the data is in the appropriate format for training the machine learning mode. This may involve steps such as data cleaning, scaling etc.

2.2.1 Feature Selection

The next step is featuring selection, where relevant features are identified for training the model. The feature selection process is crucial for training an effective machine learning model. It involves identifying the most relevant features that contribute significantly to predicting the transformer failure while reducing dimensionality and eliminating noise or irrelevant information. Some feature selection methods apply in this study were: -

2.2.1. (a) Mutual Information

In this study, we utilized mutual information to assess relationship between target variable, which denotes transformer failure, and the feature variables. The mutual dependence between two random variables that are sampled simultaneously is quantified by mutual information. Its value is always a non-negative integer, indicating the strength of the relationship between the target variable and the feature variable in predicting transformer failure.

$$I(X, y) = H(X) - H\left(\frac{X}{y}\right)$$

2.2.1. (b) Multicollinearity

It refers to existence when two or more of the predictors (independence features) moderately or highly correlated with one another. This study assessed the multicollinearity by calculating variance inflation factor (VIF) for each independent feature. Multiple machine learning models were training to evaluate the effect of removing features which had high variance inflation factor.

$$VIF = \frac{1}{1 - R^2}$$

where R – Correlation coefficient between 2 predictors

2.2.1. (c) ANOVA

In this study, ANOVA was employed as a feature selection test to identify significance of numerical features in detecting transformer faults. The ANOVA F-test is a statistical test that assesses whether the variance of two or more groups are significantly different from each other [8]. The null hypothesis assumes that the variance of the numerical feature was equal for both categories of the target variable. Conversely, the alternative hypothesis suggests that the variance of the numerical feature differed significantly between the two groups.

2.2.2 Resampling Technique

Upon analyzing the target variable, it was observed that it exhibited an imbalance. Specifically, the dataset contained 15,152 instances labeled as no fault and 5,313 instances labeled as fault. The presence of class imbalance in the target variable can introduce challenges in the modeling and evaluation process. The model may have a higher tendency to predict the majority class more frequently. This bias arises because the model's learning algorithm is naturally inclined to prioritize the majority class due to its higher prevalence in the training data. It can even completely ignore minority class in extreme cases thus leading to lead to a higher false negative rate, meaning instances of the *failure* may be misclassified as the *no failure*. After acknowledging the impact of class imbalance on the modeling process various resampling methods were explored to address this issue. Specifically, we implemented the following resampling techniques to create balanced training datasets:

2.2.2. (a) Tomek Links Under-Sampling

This is an under-sampling method that finds the most similar pairings of instances from various classes. Tomek Links seeks to strengthen the division between the classes and lessen the dominance of the majority class by eliminating the instances from the majority class in these pairs [10].

2.2.2. (b) ADASYN (Adaptive Synthetic Sampling) Over-Sampling:

This technique uses the distribution of the current examples to create synthetic cases for the minority class. It focuses on producing more synthetic instances for the minority class in hard-to-learn locations in order to more effectively address the unequal distribution. [11].

2.2.2. (c) SMOTEENN (Synthetic Minority Over-Sampling Technique Followed by Edited Nearest Neighbors)

SMOTEENN is a combined resampling technique that first applies SMOTE to generate synthetic instances for the minority class and then uses the Edited Nearest Neighbors (ENN) algorithm to remove instances that are misclassified by the nearest neighbor's classifier [12].

2.2.2. (d) (Synthetic Minority Over-Sampling Technique) and Tomek Links Under-Sampling SMOTETomek

It is a combined resampling method that integrates the SMOTE. It aims to create a more balanced representation of both the majority and minority classes in the training data. This involved selecting an instance from the minority class and interpolating between it and its k nearest neighbors from the minority class. The process was repeated until a desired balance between the majority and minority classes was achieved. Tomek Links identifies pairs of instances from different classes that are closest to each other and removes the instances from the majority class in these pairs [13].

2.2.3 Scaling Data

After finalizing the resampling technique, the study proceeded to explore different scaling techniques to preprocess the data. Four scaling techniques were employed in the experimentation:

2.2.3.(a) Robust Scaler

The Robust Scaler technique scales the features that are robust to the presence of outliers, such as the median and interquartile range.

$$RobustScalar = \frac{x - median}{IQR}$$

2.2.3. (b) Standard Scaler

The features' zero mean and unit variance are guaranteed by the Standard Scaler transformation.

$$StandardScalar = \frac{x - mean}{std\_dev}$$

2.2.3. (c) Min-Max Scaler

This technique ensures that features are scaled to a certain range, often between 0 and 1.

$$Min - \text{Max scalar} = \frac{x - min}{max - min}$$

2..2.2. (d) No Scaling

This technique involved no scaling of the feature variables. The data remained in its original form without any normalization or standardization.

2.3 Experiments

In this research study, extensive training was conducted with evaluation of multiple classification models, including AdaBoost Classifier, CatBoost Classifier, Decision Tree, Gradient Boosting, K-Neighbors Classifier, Logistic Regression, Random Forest, and XGBoost Classifier. These models were trained repeatedly, but with variations in the preprocessing techniques mentioned above. Specifically, each model was trained on different combinations of preprocessing techniques, such as scaling and resampling methods. For instance, we trained the models using SMOTE resampling technique, then repeated the training process using the SMOTEENN resampling technique, and performed the same for all the other models. Each model's performance is assessed using the proper evaluation metrics, such as area under the curve (AOC), recall, accuracy, precision, and F1-score. This evaluation provides insights into the effectiveness of each preprocessing technique and algorithm in accurately predicting transformer failure.

3 Research Elaborations

3.1 Discussion of Pre-processing Experiments

The results of the mutual information for feature selection test are presented in Table 1, the evaluation of mutual information aimed to assess the relationship between each feature and the prediction of transformer failure. Higher values of mutual information indicate a stronger dependence and contribution of the feature in predicting the target variable. Based on the initial analysis, it was observed that the feature *INUT* exhibited a relatively low mutual information score compared to the other features. This suggested that *INUT* feature had a weaker association with the prediction of transformer failure. Consequently, in our initial experimentation, we decided to remove the *INUT* feature from the dataset.

The results calculating multicollinearity with VIF are presented in Table 2. After thorough analysis and experimentation with different models, a decision was made to remove four features, namely *VL1*, *VL12*, *VL3*, and *VL23*, from the dataset. The evaluation metrics reveals that the presence or absence of these features has negligible effect on model's performance. Despite their removal, the evaluation matrices demonstrated comparable results, indicating that these features did not contribute significantly to the predictive power of the models. This step allowed us to streamline the feature space and focus on the more influential and informative predictors.

Table 1. Mutual information scores between features and target

Parameter	Mutual Info. Score
OLI	0.331134
VL12	0.238529
VL23	0.235106
VL31	0.234268
VL1	0.211439
VL2	0.207424
VL3	0.190881
IL2	0.139048
IL1	0.099537
IL3	0.097030
ATI	0.081274
OTI	0.064920
INUT	0.045288

Table 2. Variance inflation factor

Feature	VIF
OTI	14.498200
ATI	52.955866
OLI	23.600893
VL1	102439.545881
VL2	19742.716135
VL3	103896.040629
IL1	48.451082
IL2	50.312691
IL3	94.577779
VL12	41889.886053
VL23	37499.505182
VL31	15739.074408
INUT	15.399122

VIF of different features after removing VL1, VL12, VL3, and VL23 is in Table 3. It was observed that there were additional features in the dataset that exhibited multicollinearity. However, when attempts were made to remove these features, it negatively

Table 3. Variance inflation factor after removing four features

Feature	VIF
OTI	14.446010
ATI	51.288823
OLI	16.812247
VL2	69.941038
IL1	41.451000
IL2	49.527804
IL3	86.553619
VL31	24.700215
INUT	15.231403

Table 4. ANOVA f test result

Feature	f-statistic
OLI	14674.04386
VL1	6125.643956
VL3	5362.23223
VL2	5023.458076
VL12	1721.52608
VL31	1703.178606
VL23	1663.85629
ATI	1453.569741
IL2	904.826789
INUT	766.740551
OTI	567.474684
IL1	156.478289
IL3	123.110609

affected the evaluation metrics of the models. Despite the presence of multicollinearity, these features were deemed valuable for the predictive performance of the models. It is worth noting that multicollinearity is not always a detrimental factor, especially if the features still contribute meaningful information to the models. Like in this case, the models benefited from the inclusion of these features, and their removal resulted in a decline in the evaluation metrics. Therefore, considering the overall performance and evaluation results, the decision was made to retain other features which showed high multicollinearity in the analysis.

The results of the ANOVA f test presented in Table 4. The null hypothesis, assumes equal variances for both categories of the target variable, it was tested against the alternative hypothesis stating that there is significant difference in variances between the two categories. With a confidence level of 95% (significance level of 5%), the ANOVA f-test evaluates the contribution of each feature in detecting the transformer failure. Based on the results presented in Table k, we found compelling evidence to reject the null hypothesis for all features. This indicates that each feature plays a significant role in predicting transformer faults. Notably, features such as *OLI* and *VL1* exhibited the highest F-statistics, indicating the strongest statistical significance in relation to the target variable.

While the initial assessment based on mutual information suggested that *INUT* had a lower contribution to predicting transformer failure, the ANOVA test provided contrasting evidence. The high F-statistics score implied that the variance of *INUT* across different categories of the target variable was significantly different. This finding indicated that *INUT* was a valuable feature in distinguishing between various states of transformer failure. As a result, it was decided not to remove this feature from the analysis.

After conducting thorough experimentation and evaluation, we assessed the performance of various resampling methods for addressing the class imbalance in the target variable. The results, presented in Table 7, provide insights into the effectiveness of each resampling technique. Among the resampling methods tested ranged from under-sampling, over sampling and over-sampling followed by under-sampling. The SMOTEENN resampling technique, which is type of over-sampling with under-sampling outperformed other techniques in terms of *precision, recall, accuracy, F1 score and area under the curve (AUC)*. By combining the advantages of both over-sampling and under-sampling, SMOTEENN achieved a balanced representation of the classes. After finalizing the resampling technique, the study proceeded to explore different scaling techniques to preprocess the data. The objective was to evaluate the impact of scaling on the performance of the machine learning models. After conducting extensive experimentation and evaluation of different scaling techniques, the performance of each method was assessed and compared. The results, presented in Table l, shed light on the impact of scaling on the performance of the models. Among these methods, it was observed that the approach in which there was *no scaling* implemented yielded slightly better results compared to the *Min-Max Scaler technique*. Although the improvement in performance was modest, it was deemed statistically significant. The evaluation metrics, including *precision, recall, accuracy, F1 score and area under the curve (AUC)*, consistently demonstrated slightly better results for this approach. These results, which were obtained thorough experimentation and analysis, offer important new understandings into the best methods for transformer problem identification.

3.2 Model Evaluation

This section aims to provide a detailed analysis of each model with the selected preprocessing techniques using various evaluation metrics like confusion matrices and ROC curves, to assess the performance of the model in detecting transformer faults.

Confusion Matrix: The confusion matrix provides a detailed breakdown of the model's predictions and their alignment with the actual class labels. It consists of four components: true positive (TP), true negative (TN), false positive (FP), and false negative (FN). From the confusion matrix, we can calculate key metrics such as accuracy, precision, recall (sensitivity), and F1 score. Consider Failure: F, No Failure: NF, Actual Label: AL, Predicated Label: PL.

Table 6(a). Cat Boosting Classifier

CBC	Predicted Label		
		F	NF
AL	F	2563	47
	NF	1	2746

Table 6(b). Random Forest

RF	Predicted Label		
		F	NF
AL	F	2550	60
	NF	0	2747

Table 6(c). Random Forest

DFC	Predicted Label		
		F	NF
AL	F	2567	43
	NF	33	2714

Table 6(d). Random Forest

GBC	Predicted Label		
		F	NF
AL	F	2507	103
	NF	5	2742

ROC Curve: This metric illustrates the performance of different classification models. The area under this curve is used to assess the classifier's ability to correctly classify transformer failure and no failure. A higher AUC indicates better discriminatory power.

Table 6(e). AdaBoost Classifier

ABC	Predicted Label		
		F	NF
AL	F	2508	102
	NF	6	2741

Table 6(f). XGBoost

XGB	Predicted Label		
		F	NF
AL	F	2562	48
	NF	1	2746

Table 6(g). K-nearest neighbours

KNN	Predicted Label		
		F	NF
AL	F	2550	60
	NF	5	2742

Table 6(h). Logistic Repression

LR	Predicted Label		
		F	NF
AL	F	2423	187
	NF	1	2746

In final analysis of the models, we found that the *CatBoosting classifier* consistently outperformed other models in while considering *test accuracy, F1 score, and AOC (Area Under Curve)*. It achieved the highest test accuracy among all models, indicating its ability to make accurate predictions on the test dataset. Additionally, the *CatBoosting classifier* demonstrated a high *F1 score*, which is a balanced measure of precision and recall, indicating its effectiveness in capturing both the true positives and true negatives.

On the other hand, the *Decision Tree model* exhibited the highest precision among all models. This suggests that it was able to correctly classify a significant portion of positive cases as true positives, reducing the number of false positives.

While the *Random Forest* model achieved a recall of 1.0, meaning it correctly identified all instances of transformer faults as seen in Table 6, this exceptionally high recall value may raise concerns of potential overfitting to the training data. Taking into consideration both the precision and recall metrics, the *CatBoosting classifier* still stood out with a high F1 score of 0.99. This indicates a balanced performance in correctly identifying positive cases while minimizing false positives.

3.3 Explainable AI Insights

To enhance the interpretability and transparency of the predictive models, explainable AI technique shap is used in this study to gain insights and study contribution of each feature in model's prediction. Aim of this section is to get a deeper understanding of the factors influencing transformer failures and helped uncover the relationships between the features and the target variables (Fig. 1).

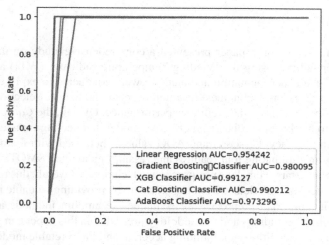

Fig. 1. ROC Curve

Figure 2 is representation of SHAP analysis. The negative SHAP values indicated a higher contribution towards failure, while positive SHAP values indicated a lower contribution towards failure. The features are arranged from bottom to top based on their feature importance. Among the features, it was observed that the OLI (Oil Level Indicator) had a significant negative impact on the prediction of failures. This finding is consistent with the results obtained from the ANOVA F-test and mutual information analysis, which indicated that the OLI feature played a crucial role in predicting transformer failures.

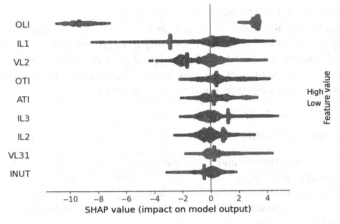

Fig. 2. SHAP analysis

4 Results

In conclusion, this research paper presented a comprehensive study on the detection of transformer failures using the Winding Temperature Indicator (WTI) as the target feature. Extensive experimentation and analysis were conducted to explore the impact of various preprocessing techniques, resampling methods, feature selection tests, and machine learning models on the predictive performance. Overall, the *CatBoosting classifier* proved to be the most robust and effective model in detecting transformer faults, achieving high accuracy, F1 score, and AOC values when used with feature selection which resulted in eliminating four features and resampling method SMOTEENN which was used to balance target labels as pre-processing techniques. Overall, this research contributes to the field of fault detection in transformers by providing valuable insights into the selection of appropriate preprocessing techniques, resampling methods, and machine learning models. The findings and methodology presented in this paper can guide future research and assist practitioners in building accurate and interpretable models for identifying transformer failures using the Winding Temperature Indicator (WTI) or similar target features present in the dataset (Table 8(a) and 8(b)).

Table 7(a). Table Experimental results of various resampling techniques

Re-Sampling	Model Name	Accuracy	F1 Score	AOC	Recall	ROC
NONE	Logistic Regression	0.849011	0.753785	0.650619	0.895833	0.864282
	AdaBoost Classifier	0.898607	0.810069	0.78388	0.838068	0.878863

(*continued*)

Table 7(a). (*continued*)

Re-Sampling	Model Name	Accuracy	F1 Score	AOC	Recall	ROC
	Gradient Boosting	0.892499	0.811644	0.740625	0.897727	0.894204
	K-Neighbors Classifier	0.915221	0.83883	0.823154	0.855114	0.895617
	Decision Tree	0.914732	0.849244	0.780778	0.930871	0.919996
	CatBoosting Classifier	0.935744	0.879963	0.849339	0.912879	0.928287
	Random Forest	0.937454	0.88353	0.850263	0.919508	0.931601
	XGBClassifier	0.937699	0.884354	0.848564	0.923295	0.933001
ADASYN Over-sampling	Logistic Regression	0.904116	0.913663	0.84195	0.998729	0.90258
	AdaBoost Classifier	0.922195	0.928804	0.867789	0.999047	0.920947
	Gradient Boosting	0.928006	0.93367	0.877551	0.997458	0.926879
	K-Neighbors Classifier	0.935432	0.936568	0.934789	0.938354	0.935384
	Decision Tree	0.93301	0.938124	0.883708	0.999682	0.931928
	CatBoosting Classifier	0.954641	0.956882	0.925223	0.990785	0.954054
	Random Forest	0.956255	0.958069	0.933655	0.983794	0.955808
	XGBClassifier	0.956416	0.958766	0.922964	0.997458	0.95575
Over-sampling followed by Under-sampling	Logistic Regression	0.893148	0.901679	0.822792	0.997298	0.894931
	AdaBoost Classifier	0.918865	0.923486	0.86035	0.996623	0.920197
	Gradient Boosting	0.92517	0.929077	0.869335	0.997636	0.926411
	K-Neighbors Classifier	0.930148	0.932672	0.885784	0.984802	0.931084
	Decision Tree	0.93778	0.93707	0.931288	0.942925	0.937868
	CatBoosting Classifier	0.955036	0.955654	0.926984	0.986153	0.955569
	Random Forest	0.956529	0.957287	0.925307	0.991557	0.957129
	XGBClassifier	0.957193	0.957719	0.930277	0.986829	0.9577

Table 7(b). Table Experimental results of various resampling techniques

Re-Sampling	Model Name	Accuracy	F1 Score	AOC	Recall	ROC
Over-sampling followed by Under-sampling (SMOTE-ENN)	Logistic Regression	0.964906	0.966901	0.936243	0.999636	0.963994
	AdaBoost Classifier	0.979839	0.98068	0.964122	0.997816	0.979368
	Gradient Boosting	0.979839	0.980687	0.963796	0.99818	0.979358
	K-Neighbors Classifier	0.986	0.986346	0.986526	0.986167	0.985995
	Decision Tree	0.987866	0.988286	0.978587	0.99818	0.987596
	CatBoosting Classifier	0.988613	0.989019	0.978276	1	0.988314
	Random Forest	0.990853	0.991157	0.98282	0.999636	0.990623
	XGBClassifier	0.99104	0.991336	0.983172	0.999636	0.990814
SMOTE Over-sampling	Logistic Regression	0.895067	0.904072	0.830424	0.992056	0.89537
	AdaBoost Classifier	0.91932	0.924873	0.862959	0.996359	0.919561
	Gradient Boosting	0.92493	0.929773	0.871024	0.997021	0.925155
	K-Neighbors Classifier	0.92757	0.931116	0.885143	0.982125	0.92774
	Decision Tree	0.934169	0.934687	0.924547	0.945051	0.934203
	CatBoosting Classifier	0.954133	0.955732	0.920835	0.99338	0.954256
	Random Forest	0.955288	0.956689	0.924907	0.990732	0.955399
	XGBClassifier	0.956113	0.957138	0.932496	0.983118	0.956197
Tomek Links Under-sampling	Logistic Regression	0.864513	0.780349	0.670621	0.93301	0.886856
	AdaBoost Classifier	0.913849	0.836502	0.819367	0.854369	0.894447
	Gradient Boosting	0.909842	0.842244	0.767572	0.93301	0.917399
	K-Neighbors Classifier	0.920361	0.846228	0.842967	0.849515	0.897251

(*continued*)

Table 7(b). (*continued*)

Re-Sampling	Model Name	Accuracy	F1 Score	AOC	Recall	ROC
	Decision Tree	0.91986	0.857143	0.793388	0.932039	0.923832
	CatBoosting Classifier	0.944904	0.897004	0.866184	0.930097	0.940074
	Random Forest	0.946156	0.899953	0.864164	0.938835	0.943768
	XGBClassifier	0.94866	0.904518	0.869293	0.942718	0.946722

Table 8(a). Table Experimental results of various scaling techniques

Scaling	Model Name	Accuracy	F1 Score	AOC	Recall	ROC
None	CatBoosting Classifier	0.99104	0.991336	0.983172	0.999636	0.990814
	XGBClassifier	0.990853	0.991157	0.98282	0.999636	0.990623
	Random Forest	0.9888	0.989197	0.978625	1	0.988506
	K-Neighbors Classifier	0.987866	0.988286	0.978587	0.99818	0.987596
	Decision Tree	0.985813	0.986192	0.984403	0.987987	0.985756
	AdaBoost Classifier	0.979839	0.98068	0.964122	0.997816	0.979368
	Gradient Boosting	0.979839	0.980687	0.963796	0.99818	0.979358
	Logistic Regression	0.964906	0.966901	0.936243	0.999636	0.963994
Min Max Scaler	CatBoosting Classifier	0.990894	0.991182	0.98322	0.999274	0.990685
	XGBClassifier	0.990151	0.990459	0.982851	0.998186	0.98995
	Random Forest	0.988664	0.989054	0.978346	1	0.988381
	K-Neighbors Classifier	0.987549	0.987965	0.9783	0.997823	0.987292
	Gradient Boosting	0.980301	0.981105	0.964261	0.998549	0.979846
	AdaBoost Classifier	0.978814	0.97965	0.96416	0.995646	0.978394
	Decision Tree	0.978071	0.978723	0.97276	0.984761	0.977904
	Logistic Regression	0.957629	0.960279	0.923592	1	0.956571

(*continued*)

Table 8(a). (*continued*)

Scaling	Model Name	Accuracy	F1 Score	AOC	Recall	ROC
Robust Scaler	XGBClassifier	0.986101	0.986557	0.975541	0.997825	0.985834
	K-Neighbors Classifier	0.986101	0.986576	0.974196	0.999275	0.985801
	CatBoosting Classifier	0.984989	0.985502	0.973135	0.998187	0.984689
	Random Forest	0.984062	0.984632	0.970754	0.998912	0.983725
	Decision Tree	0.981468	0.981936	0.978402	0.985497	0.981376
	Gradient Boosting	0.970904	0.972266	0.947985	0.997825	0.970292
	AdaBoost Classifier	0.969051	0.970531	0.945342	0.997099	0.968413
	Logistic Regression	0.952372	0.955467	0.915035	0.999637	0.951297

Table 8(b). Table Experimental results of various scaling techniques

Scaling	Model Name	Accuracy	F1 Score	AOC	Recall	ROC
Standard Scaler	CatBoosting Classifier	0.987796	0.988327	0.978634	0.998214	0.987417
	XGBClassifier	0.987611	0.988131	0.979972	0.996427	0.98729
	Random Forest	0.985947	0.986582	0.975218	0.998214	0.9855
	K-Neighbors Classifier	0.985577	0.986224	0.975201	0.997499	0.985143
	Decision Tree	0.976701	0.977484	0.977833	0.977135	0.976685
	Gradient Boosting	0.975037	0.976403	0.955852	0.997856	0.974206
	AdaBoost Classifier	0.972078	0.973578	0.954047	0.993926	0.971283
	Logistic Regression	0.962833	0.965327	0.933289	0.999643	0.961493

References

1. Lei, Y.: Intelligent fault diagnosis and remaining useful life prediction of rotating machinery. Butterworth-Heinemann (2016)
2. Lei, Y., Jia, F., Lin, J., Xing, S., Ding, S.X.: An intelligent fault diagnosis method using unsupervised feature learning towards mechanical big data. IEEE Trans. Industr. Electron. **63**(5), 3137–3147 (2016). https://doi.org/10.1109/TIE.2016.2519325

3. Saha, T.K., Purkait, P.: Investigation of polarisation and depolarisation current measurements for the assessment of oil-paper insulation of aged transformers. IEEE Trans. Dielectr. Electr. Insul. **11**, 144–154 (2004)
4. Blennow, J., Ekanayake, C., Walczak, K., Garcia, B., Gubanski, S.M.: Field experiences with measurements of dielectric response in frequency domain for power transformer diagnostics. IEEE Trans. Power Delivery **21**(2), 681–688 (2006)
5. IEEE Guide for the Interpretation of Gases Generated in Oil-Immersed Transformers. IEEE Standard C57.104-2008 (2009)
6. Balan, A., et al.: Detection and analysis of faults in transformer using machine learning. In: Proceedings of the 2023 International Conference on Intelligent Data Communication Technologies and Internet of Things (IDCIoT), Bengaluru, India, pp. 477–482 (2023). https://doi.org/10.1109/IDCIoT56793.2023.10052786
7. Pias, T.S., Su, Y., Tang, X., Wang, H., Yao, D.(D.): Undersampling for fairness: achieving more equitable predictions in diabetes and prediabetes. Cold Spring Harbor Laboratory (2023)
8. Elssied, N.O.F., Ibrahim, O., Osman, A.H.: A novel feature selection based on one-way ANOVA f-test for e-mail spam classification. Res. J. Appl. Sci. Eng. Technol. **7**(3), 625–638 (2014)
9. Johnson, J.M., Khoshgoftaar, T.M.: Survey on deep learning with class imbalance. J. Big Data **6**, 27 (2019). https://doi.org/10.1186/s40537-019-0192-5
10. Elhassan, T., Aljourf, M., Al-Mohanna, F., Shoukri, M.: Classification of imbalance data using tomek link (T-Link) combined with random under-sampling (RUS) as a data reduction method. Glob. J. Technol. Optim. **1** (2016). https://doi.org/10.4172/2229-8711.S1111
11. Hu, S., Liang, Y., Ma, L., He, Y.: MSMOTE: Improving classification performance when training data is imbalanced. In: Proceedings of the 2009 Second International Workshop on Computer Science and Engineering (2009)
12. Liu, D., Zhang, H., Polycarpou, M.M., Alippi, C., He, H. (eds.) Advances in Neural Networks - ISNN 2011 – Proceedings of the 8th International Symposium on Neural Networks, ISNN 2011, Guilin, China, 29 May–1 June 2011, Part III. Lecture Notes in Computer Science, vol. 6677. Springer (2011)
13. Karvelis, P., Spilka, J., Georgoulas, G., Chudáček, V., Stylios, C.D., Lhotská, L.:. Combining latent class analysis labeling with multiclass approach for fetal heart rate categorization. Physiol. Meas. (2015)
14. Attenberg, J., Ertekin, Ş.: Class Imbalance and Active Learning. Wiley (2013)

Type 2 Diabetes Prediction Using Machine Learning: A Game-Changer for Healthcare

Sadhana Singh[✉], Priyanka Sharma, and Pragya Pandey

CSE-AI, ABESIT, Ghaziabad, India
{sadhana.singh,priyanka.sharma,pragya.pandey}@abesit.edu.in

Abstract. In today's scenario, Diabetes is the most common disease for human beings. In diabetes, the sugar level is increased due to the genetic problem or due to tension taken by human beings. The main aim of this study is to predict the diabetes by using Machine learning techniques. In this paper we collect the data which is affected be diabetes using the characteristics defined by American Diabetes Association (ADA) criteria. We evaluate the real-world data by using machine learning techniques. In this, we obtain the precision value is less than 5.7, then the person is non-diabetic and if the precision value is greater than or equal to 6.5, then the person is diabetic. This precision value is achieved with the Glycosylated Hemoglobin test.

Keywords: Type 2 Diabetes · Machine Learning · American Diabetes Association · glycosylated hemoglobin test

1 Introduction

In this paper we have to perform diabetic prediction by using the Machine Learning Technique. Diabetes is that which occurs due to tension or today's lifestyle of human beings. In diabetes, human beings have different problems like eye sight weak, kidney problem, heart problem, high blood pressure or cholesterol problem, hearing problem, etc. Diabetes is spread due to the spreading the tension in our mind.

Symptoms of Diabetes like time-to-time drinking water due to thrust. Many times, urine is passing very rapidly. This is also a very serious problem for diabetic patients. Diabetes is the metabolic issue of an individual who has high sugar levels or high glucose levels in the human body. This happens due to the loss of insulin made by the pancreas in the human body. Diabetes is of mainly two types: Type1 Diabetes and Type2 Diabetes. In type1 diabetes, the pancreas does not produce the insulin to the human body and the individual takes the insulin injection or insulin pump for generating the insulin in the body. In type2 diabetes, the pancreas produces insulin but the body cannot use it effectively. In today's life, type2 diabetes is very common. Use the enter key to start a new paragraph. The appropriate spacing and indent are automatically applied.

Diabetes is a long-lasting sickness that happens your description accurately captures the essence of diabetes, a chronic condition that significantly impacts the body's ability to regulate blood glucose levels. Diabetes is a chronic condition characterized by a dysfunction in insulin production or the body's inability to effectively use insulin. Insulin is a hormone produced by the pancreas, and it plays a crucial role in regulating blood glucose (sugar) levels. Insulin acts as a regulator, helping to normalize blood glucose levels by facilitating the uptake of glucose into cells for energy.

American Diabetes Association are: (1) a level of glycated haemoglobin (HbA1c) greater or equal to 6.5%; (2) basal fasting blood glucose level greater than 126 mg/dL, and; (3) blood glucose level greater or equal to 200 mg/dL 2 h after an oral glucose tolerance test with 75 g of glucose [2].

The type 2 diabetes that Covid 2019 may cause could be severe. The Python programme is used to differentiate between normal sugar patient and type 2 diabetic patient in the current study. We are collecting the data from the dataset. Absolutely, the Diabetes prediction dataset you described is valuable for building machine learning models aimed at predicting diabetes based on a patient's medical and demographic information. Factors such as BMI, hypertension, heart disease, smoking history, HbA1c level, and blood glucose level offer insights into the patient's health status. The Diabetes prediction dataset provides a foundation for developing machine learning models that contribute to early detection, personalized treatment plans, and improved patient outcomes in the context of diabetes. It aligns with the broader trend of leveraging data and analytics in healthcare for more proactive and personalized patient care.

2 Literature Review

Quan Zou, Kaiyang Qu, Yamei Luo, Dehui Yin, Ying Ju and Hua Tang in 2018: Methodology: Compared the accuracy of Principal Component Analysis (PCA), using all features, and using minimum Redundancy Maximum Relevance (mRMR) for diabetes prediction. Findings: Results indicated that using all features and mRMR had better accuracy than PCA. Fasting glucose alone showed better performance, especially in the Luzhou dataset [3]. Umair Muneer Butt, Sukumar Letchmunan, Mubashir Ali, Fadratul Hafinaz Hassan, Anees Baqir and Hafiz Husnain Raza Sherazi in 2021: Methodology: Proposed an MLP-based algorithm for diabetes classification and a deep learning-based Long Short-Term Memory (LSTM) model for diabetes prediction. Also, introduced an IoT-based real-time diabetic monitoring system using a smartphone, BLE-based sensor device, and machine learning for predicting blood glucose (BG) levels. Contribution: Provided a comprehensive approach combining machine learning, deep learning, and IoT for diabetes management [4]. Francesco Mercaldo, Vittoria Nardone and Antonella Santone in 2017: Methodology: Proposed a machine learning algorithm to discriminate between diabetes-affected and non-affected patients. Evaluated the method on real-world data from the Pima Indian population near Phoenix, Arizona. Objective: Aimed to develop a method for effective diabetes classification using machine learning techniques [5]. Muhammad Exell Febrian, Fransiskus Xaverius Ferdinan, Gustian Paul Sendani, Kristein Margi Suryanigrum and Rezki Yunanda in 2023: Methodology: Proposed two k-Nearest Neighbor algorithms and the Naive Bayes algorithm for predicting diabetes based on health attributes in the dataset using supervised machine learning. Approach: Employed traditional machine learning algorithms for diabetes prediction [6]. Aishwarya Mujumdar and Dr. Vaidehi V in 2019 Methodology: Proposed a diabetes prediction model for better classification of diabetes. Objective: Aimed at improving the accuracy of diabetes prediction through the development of a novel prediction model [6]. These studies collectively highlight the diverse approaches and methodologies researchers have employed to address the challenge of diabetes prediction. The integration of machine learning, deep learning, and IoT in some studies reflects the interdisciplinary nature of efforts to improve diabetes management and prediction. Each study contributes valuable insights to the field of healthcare and diabetes research.

3 Research Methodology

In research methodology we have to collect the data on the dataset.

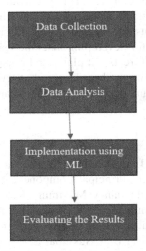

Fig. 1. Research Framework

Figure 1 shows how to find out the result. In the data collection process, we have to collect the data from the dataset which are provided on the Kaggle website. Then we have to analyse the data on the basis of the parameters defined by the American Diabetes Association group. We have to implement these on the Python language with the help of the Pandas, Matplotlib and seaborn library. Evaluating the results with the help of these datasets and Python libraries.

4 Proposed Work

In this paper, we have to perform different machine learning techniques for diabetes prediction (Fig. 2).

In supervised learning, one person has to guide the data. Supervised learning techniques are used to construct the predictive model for diabetes. Predictive models are used to predict the data with the help of the dataset. Decision trees, Bayesian Network,

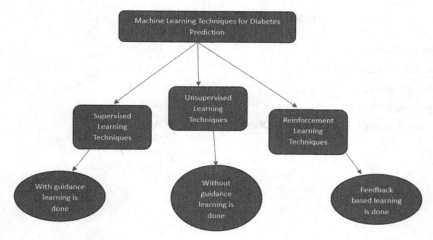

Fig. 2. Proposed Work for Diabetes Prediction

Artificial Neural Network and Instance based learning techniques are the examples of the supervised learning techniques. In this learning the inputs and outputs of the problem are known.

In unsupervised learning, there is no proper guidance for data. In this learning technique inputs are known but the outputs are unknown. This technique is basically used for the transactional data. K-means and K-medians techniques are the examples of the unsupervised learning techniques.

In the reinforcement learning, we know the output and input also but output is work like input for the perspective of the data or problem.

In this paper we simply use the Decision tree machine learning algorithm for calculating the diabetes prediction level with the help of the ADA standard for diabetes.

5 Results and Discussion

For experiments we have to take the dataset from the Biostatistics program at Vanderbilt, African Americans in Virginia. This dataset is designed in the 2023. In this paper we refer the diagnosis process for diabetes patient is based on the Glycosylated Heomoglobin (HbA1c) test parameters.

Table 1 shows the parameters for finding the patient is diabetic or not. Target goals of <7.0 % may be beneficial for the diabetes patients.

For experiment we have to collect the data about 390 persons. In this experiment 163 males and 228 females. We have to collect the data through the dataset on the parameters like, Cholesterol, Glucose, Age, Gender, Height, Weight, BMI (Body Mass Index), HDL (High Density Lipoproteins), BP (Blood Pressure). HDL is the type of good cholesterol.

Figure 3 shows the age of different patients which we are taken for the process of diabetes test. Figure 4 shows the how to take the glucose of different patients of test. Figure 5 shows the glucose level of patient ifs decade on the basis of average glucose

Table 1. Configuration of HBA1c Test

Reference Group	HbA1c in %
Non-Diabetic adults>=18	<5.7
At risk (Prediabetes)	5.7-6.4
Diagnosing Diabetes	>=6.5
Therapeutic goals for controlling the Diabetes	Age>19 Goal of Therapy <7.0 Age<19 Goal of Therapy <7.5

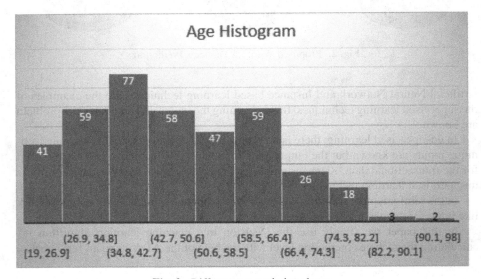

Fig. 3. Different age variation shows.

level is lie between 5.7–6.4. Figure 6 shows the decade of overall glucose level on the basis of HbA1c test.

In Fig. 7 we have to take the 5 patients glucose level to shows the different results. Figure 8 we shows the 5 patients results on the basis of different parameters like fasting insulin, fasting glucose, converted glucose, HOMA (this the index for calculating the fasting glucose and level of insulin taken by the patients), hemoglobin level.

Figure 9 shows the number of patients in terms of female and male for performing the test. Figure 10 shows the calculation regarding BMI (Body Mass Index) by gender. Figure 11 shows the calculation of sum of the waist or hip with respect to the different male and female patients. Figure 12 shows the sum of glucose on different genders. Figure 13 shows the sum of cholesterol level of different patients on the basis of gender.

Type 2 Diabetes Prediction Using Machine Learning 39

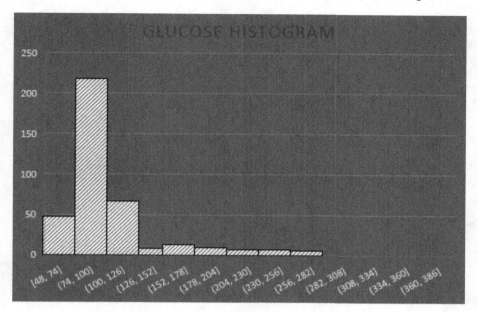

Fig. 4. Different glucose patterns for patients

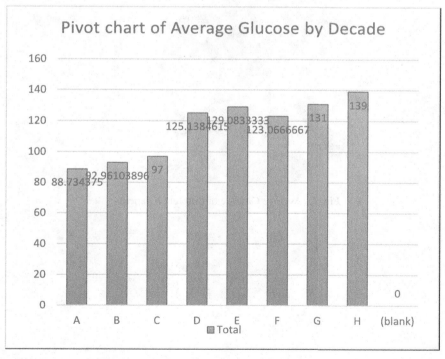

Fig. 5. Average Glucose level is decade for patients

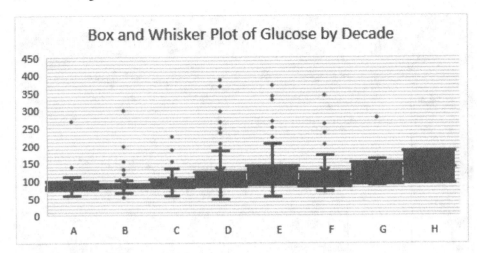

Fig. 6. Glucose level is decade for patients

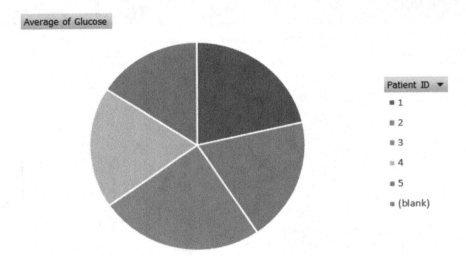

Fig. 7. Average Glucose of different 5 patients

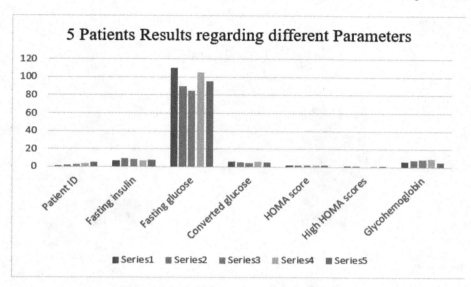

Fig. 8. Diabetes calculated on different parameters

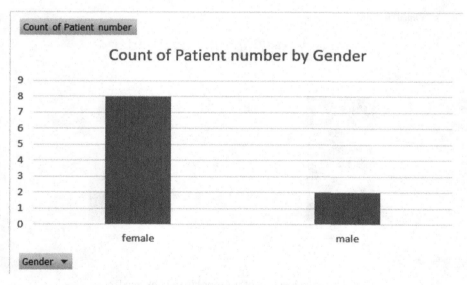

Fig. 9. Calculate number of patients

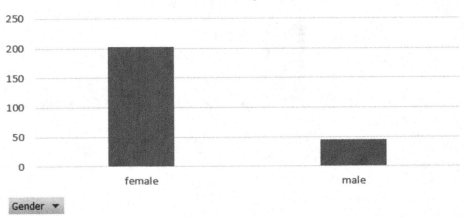

Fig. 10. Calculate sum of BMI by gender

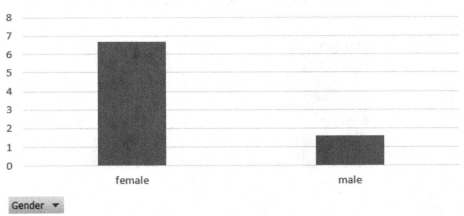

Fig. 11. Calculate sum of waist or hip ratio by gender

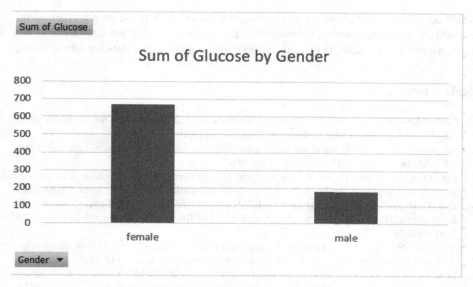

Fig. 12. Calculate sum of glucose by gender

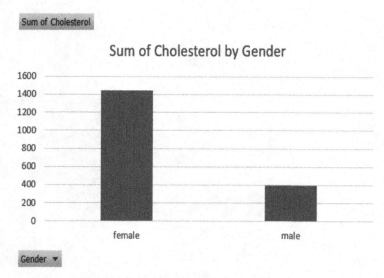

Fig. 13. Calculate sum of cholesterol by gender

6 Conclusion

In this paper we simply focus on the type 2 diabetes. In this paper, we show the different parameters to calculate the diabetes are present in patient or not. We show the different results for calculating the diabetes in patient. In this paper, we had to take the parameters for calculating the diabetes with the help of HbA1c test. In this we have to use the machine learning technique for calculating the patient is diabetic or not. We use the supervised machine learning technique for this purpose. In this we have to take data from dataset and we saw the 60 persons are diabetic and 331 patients are non-diabetic.

References

1. Global Burden of Disease Collaborative Network. Global Burden of Disease Study 2019. Results. Institute for Health Metrics and Evaluation (2020)
2. AD Association: Classifcation and diagnosis of diabetes: standards of medical care in diabetes-2020. Diabetes Care (2019). 10. 2337/dc20-S002
3. Zou, Q., Qu, K., Luo, Y., Yin, D., Ju, Y., Tang, H.: Predicting diabetes mellitus with machine learning techniques. **9**, Article 5, November 2018. https://doi.org/10.3389/fgene.2018.00515
4. Butt, U.M., et al.: Machine learning based diabetes classification and prediction for healthcare applications. Hindawi J. Healthc. Eng. **1**, Article ID 9930985 (2021). https://doi.org/10.1155/2021/9930985
5. Mercaldo, F., Nardone, V., Santone, A.: Diabetes mellitus affected patients classification and diagnosis through machine learning techniques. Procedia Comput. Sci. **112**, 2519–2528 (2017)
6. Febrian, M.E., Ferdinan, F.X., Sendani, G.P., Suryanigrum, K.M., Yunanda, R.: Diabetes prediction using supervised machine learning. Procedia Comput. Sci. **216**, 21–30 (2023). Proceedings of the 7th ICCSCI 2022

Classification of Brain Tumors in MRI Images Using Deep Learning

Rahul Namdeo Jadhav[1,2(✉)] and G. Sudhagar[1]

[1] Department of ECE, Bharath Institute of Higher Education and Research, Chennai, Tamilnadu, India
jadhavrn@gmail.com, sudhagar.ece@bharathuniv.ac.in
[2] AISSMS Institute of Information and Technology, Pune, India

Abstract. Brain tumors are currently the biggest cause of mortality throughout the globe. This pattern has been seen in recent years. A brain tumor is characterized by abnormal cell proliferation in the brain. The purpose of this research is to develop a dependable brain tumor segmentation model based on the GoogLeNet architecture and compare its performance to AlexNet and VGGNet. Data pre-processing, Google model training, and testing on a preset dataset are all essential components of the approach. A comparison with AlexNet and VGGNet illustrates our proposed model's utility and specificity in brain tumor segmentation. The results emphasize GoogLeNet's utility by indicating potential breakthroughs in early tumor identification and classification. The study focuses on the utilization of cutting-edge medical imaging technologies, such as GoogLeNet and the Walrus Optimization Algorithm, to increase diagnosis accuracy and treatment choices for brain tumors.

Keywords: MRI images · brain tumors · deep learning · classification · walrus

1 Introduction

Although medical knowledge and biological advancements have made significant progress in the last several decades in curing many illnesses, cancer's unpredictable nature continues to be a social embarrassment. One of the scariest and fastest-growing illnesses is brain tumour cancer. The largest and most intricate organ is the human brain. The neurological system regulates a wide range of bodily processes, including breathing, muscle contraction, and sensory perception. Each cell is different, with some growing properly and others losing their ability, resisting, or developing abnormally [1]. Massive clumps of aberrant cells that combine to create tissue are called tumours. Uncontrolled and abnormal brain cell development is the cause of cancerous brain tumours. It is among the deadliest and severe cancers. Over the years, there has been a substantial advancement in the processes for assessing tumours. Though regrettably, we do not yet have the means to treat brain tumours, we are moving closer to our goal thanks to our diagnostic technologies. Brain tumours may be detected and treated using a magnetic resonance machine (MRI) as assistance [2]. Because brain tumours may be easily identified from

MRI scans, we employ them for high-quality imaging. They facilitate the prompt identification and discovery of tumour constituents. Right now, every process relies on human comprehension, which might lead to inaccurate results. It is difficult to find these tumours while they are still small at an early stage. Medical professionals cannot get exact findings from imaging procedures like CT and MRI [3]. As a result, they will wait and carry out the imaging process again. The tumour will continue to develop and endanger the person's life if they predict mistakenly that there isn't a tumour there. Compared to conventional techniques, machine-based identification and evaluation could be simpler and more lucid [4]. One way to deal with tumours that were not present when the item was registered is to use tumour-growth modelling. For this, two different tumour-growth models were investigated. A more comprehensive tumour-growing variant was more likely to provide segmentation that was meaningful and unique, even if a decreased tumour growth model was quicker to calculate. The two techniques have been integrated into a single framework for assessing brain pictures with tumours that usually makes use of all imaging data that is accessible in laboratories [5].

There does not yet exist a universally accurate method for detecting brain tumours, regardless recent research, regardless of the tumour's location, form, or intensity study on the recognition and division of brain tumours. Many algorithms for classifying brain tumours and extracting features have been reported in recent study [6]. In low-level feature extraction, grey-level co-occurrence matrices, or GLCMs, are often used. Furthermore, feature extraction methods such as Fisher Vector, Neural Network, and Bag-of-Words (BoW) are designed to address the complex structure of a brain tumour. A recent study discovered that brain cancers might be identified with a precision level between 71.39 and 94.68% by combining the fisher vector approach with efficient site pooling to discriminate between three forms of malignancies: pituitary astrocytoma, meningioma, and glioma [7]. With the help of this segmentation system, radiation evaluation, analysis, proof of identity, treatment planning, and monitoring may be enhanced. A number of the researchers looked at different approaches of classifying brain tumours; these are discussed in more detail below [8].

The use of small kernels, according to the author, allows for deeper architectural design. His research shows that CNN archives with superior accuracy and low complexity [9]. He used CNN to solve the classification challenge in his study and proposed a novel multi-layered model for categorising MRI brain tumours [10]. The experimental results demonstrate how well the suggested technique works compared to the existing methods [11].

2 Related Work

This work aims to use MRI images for brain tumour diagnosis. Deep learning networks, or CNN models, are used in diagnosis. The foundation is CNN model Resnet50. Eight more layers were added to the Resnet50 model in place of the last five levels. The accuracy of this model is 97.2%. The models that provide the findings include Googlenet, Resnet50, Dense net201, InceptionV3, and Alexnet. The most efficient model categorised brain tumour pictures. According to previous research, the discovered approach is successful and may be utilised in computerised systems to diagnose brain tumours [12].

The study's goal was to save the doctor's time by limiting the use of cranial MR scans to those with a mass. The categorization accuracy of cranial magnetic resonance imaging in the experimental trials is 97.18%. The performance of the suggested strategy outperformed that of the other recent research in the literature, according to the evaluated findings. The effectiveness of the suggested strategy and its applicability to computer-assisted brain tumour identification were further shown by the experimental findings [13].

This work classifies common brain tumours (glioma, meningioma, pituitary) using deep transfer of information and a GoogLeNet that has been trained. Through patient-level cross-validation on fig share MRI data, the proposed method beats previous algorithms with 98% mean accuracy. It provides minimal training examples and analyses misclassifications to show applicability [14].

This study provides an in-depth examination of both state-of-the-art deep learning methods and conventional machine learning approaches for the identification of brain tumours. The main accomplishments as shown by the performance evaluation metrics of the three diagnostic procedures' applicable algorithms are highlighted in this review study. Furthermore, as a guide for further research, this paper highlights the important discoveries and highlights the lessons discovered [15].

This paper presents deep learning-based brain tumour segmentation utilising several MRI modalities. The In order to avoid overfitting, a hybrid convolutional neural network (CNN) combines batch normalisation and dropout regularisation with patch-based local and context-sensitive data. Data imbalance is managed using the two-phase training approach. On the BRATS 2013 dataset, the methodology enhances dice score, empathy, and precision in the total tumour area by 0.86, 0.86, and 0.91 when compared to state-of-the-art approaches [16].

This research grades brain tumour locations using MRI analysis and Machine-Learning-Technique (MLT). Compared to noise-corrupted MRI slices, SGO-assisted Fuzzy-Tsallis thresholding improves tumour identification in pre-processing. In post-processing, Level-Set Segmentation (LSS) is confirmed using Active-Contour (ACS) and Chan-Vese (CVS). Grey Using tumour data, the Level the Combination Matrix (GLCM) chooses characteristics. SVM-RBF outperforms Random-Forest and k-Nearest Neighbour classifiers on the BRATS2015 database with >94% accuracy [17].

3 Background

This section provides background information about brain tumors and the segmentation process, along with a summary of our suggested GoogLeNet model and alternative designs like AlexNet and VGGNet. This research extensively tests Deep Convolutional Neural Networks (ConvNets) for brain tumor classification utilizing multi-sequence MR data. Novel ConvNet models trained from scratch on MRI patches, slices, and multi-planar volumetric slices are proposed. LGG categorization with/without 1p/19q scores 97% [18]. This study introduces TumorDetNet, a unified end-to-end deep learning model for brain tumor detection and classification. It identified brain tumors with 99.83% accuracy, classified benign and malignant brain tumors with 100% accuracy, and identified meningiomas, pituitary, and gliomas with 99.27% accuracy [19].

Brain Tumor Segmentation

The process of "brain tumour segmentation" involves separating tumours from other types of anomalies in brain MRI scans. Brain tumour subtypes are classified more precisely using MRI segmentation. This aids in directing the next diagnostic procedure. Planning radiation or surgical procedures requires accurate delineation, which it enables. The process of picture A common technique in brain magnetic resonance imaging analysis is segmentation. The structural features of the brain may be measured and visualised, sick areas can be characterised, surgical procedures can be planned, and therapies guided by images can be administered. Brain tumour segmentation is used to identify the location and size of tumour regions, with a focus on: active tumour cluster, dead tissue, or necrosis; and, Edoema (tumour-related swelling).

3.1 Walrus Optimization Algorithm (WaOA)

The Walrus Optimisation Algorithms (WaOA) is a metaheuristic that is population-based, with its population members represented by walruses. In WaOA, these walruses symbolize potential solutions to the optimization problem, and their positions define the specifications for issue variables in the search space. As a result, each the walrus is seen as a vector, and the whole community of the walrus is mathematically represented expressed as a population matrix. Initially, walrus populations are formed at random during the introduction of WaOA. The WaOA population matrix's construction is precisely defined using Eq. (1).

$$X = \begin{bmatrix} X_1 \\ \vdots \\ X_i \\ \vdots \\ X_N \end{bmatrix}_{N*M} = \begin{pmatrix} x_{1,1} & \cdots & x_{1,j} & \cdots & x_{1,m} \\ \vdots & \ddots & \vdots & & \vdots \\ x_{i,1} & \cdots & x_{i,j} & \cdots & x_{i,m} \\ \vdots & & \vdots & \ddots & \vdots \\ x_{N,1} & \cdots & x_{N,j} & \cdots & x_{N,m} \end{pmatrix}_{N*m} \quad (1)$$

In the given context, the population of walruses is represented as X, where each individual walrus, denoted as Xi, stands for a candidate solution. Within this framework, xi, j signifies the value proposed by the ith walrus for the jth decision variable. The population comprises N walruses, and the problem involves m decision variables. Each walrus serves as a potential solution to the issue, as well as recommended values for variables to consider allow us to calculate the objective function. The estimated objective function values resulting from the contributions of these walruses are defined in Eq. (2).

$$F = \begin{bmatrix} F_1 \\ \vdots \\ F_i \\ \vdots \\ F_N \end{bmatrix}_{N*1} = \begin{bmatrix} F(X_1) \\ \vdots \\ F(X_i) \\ \vdots \\ F(X_N) \end{bmatrix}_{N*1} \quad (2)$$

Here, F is the vector of goal operations, with each element represented as Fi, which represents the individual the desired function's value is obtained from the inputs that are provided of the its walrus (Table 1).

4 Methodology

The major goal of this research is to create an effective brain tumour segmentation model utilising the GoogLeNet architecture. We want to compare the performance of our model to that of two well-known convolutional neural network designs, AlexNet and VGGNet. The technique includes essential processes including data pre-processing, training the model using the GoogLeNet architecture, as well as subsequent testing against a test set. The comparison with AlexNet and VGGNet provides information on the efficacy & specificity of our proposed model in context of brain tumour segmentation.

Data set:
The data collected from Kaggle website:
https://www.kaggle.com/datasets/leaderandpiller/brain-tumor-segmentation image dataset. And the figure illustrates MRI images of Brain Tumour.

Dataset Description
This Kaggle dataset is intended for brain tumour segmentation, and it is divided into two folders: 'yes' (identifying the existence of a tumour) and 'no' (showing the absence of a tumour). It is an invaluable asset for training machine learning algorithms to recognise and outline brain tumours in medical photos. The dataset is critical for the development of techniques that may automate the identification as well as segmentation of brain tumours, allowing for more efficient medical imaging analysis.

4.1 Feature Selection Using Walrus Optimization Algorithm (WaOA)

For feature selection, we will include the Walrus Optimisation Algorithm. WaOA is an evolutionary algorithm that simulates walrus hunting behaviour. It may be used to detect and select significant characteristics from picture data, optimising the feature subset for higher classification accuracy. And We discussed about Walrus Optimization Algorithm in the background section in brief.

Train_test Split
In the train-test split, the dataset is separated into training and test sets. The model learns patterns and characteristics from the training set, which contains most of the data. Test the model's generalisation performance on an unseen dataset independent from training. This split tests the model with fresh data, giving a valid measure of how well it performs and identifying overfitting. To balance training sufficiency with rigorous assessment, common splits allocate 70–80% of data to training and 20–30% to testing.

Table 1. Algorithm: pseudocode of WaOA

Algorithm: pseudocode of WaOA

Start WaOA
1. Enter all the data for the optimisation issue.
2. Decide The number of walruses (N) and the total number of iterations (T)
3. Locations of walruses are initialised.
4. Considering t=1:T
5. Revise the strongest walrus according to the objective function value standard.
6. Regarding i=1:N
7. **Phase 1: Plan of feeding (examination)**
8. Find the jth walrus's new location using

$$x_{i,j}^{P_1} = x_{i,j} + rand_{i,j} \cdot (SW_j - I_{i,j} \cdot x_{i,j})$$

9. Update the ith walrus location using

$$X_i = \begin{cases} X_i^{P_1}, F_i^{P_1} < F_i, \\ X_i, else, \end{cases}$$

10. **Phase2: Migration**
11. Select where the ith walrus will be admitted.
12. Determine the walrus's new position using

$$x_{i,j}^{P_2} = \begin{cases} x_{i,j} + rand_{i,j} \cdot (x_{k,j} - I_{k,j} \cdot x_{i,j}), & F_k < F_i, \\ x_{i,j} + rand_{i,j} \cdot (x_{i,j} - x_{k,j}), & else, \end{cases}$$

13. Update the ith walrus location using

$$X_i = \begin{cases} X_i^{P_2}, F_i^{P_2} < F_i, \\ X_i, & else \end{cases}$$

14. **Phase 3: Getting away and fending off predators**
15. Determine a new location near the ith walrus using

$$x_{i,j}^{P_3} = x_{i,j} + \left(lb_{local,j}^t + \left(ub_{local,j}^t - rand \cdot lb_{local,j}^t\right)\right),$$

$$Local\ bounds: \begin{cases} lb_{local,j}^t = \dfrac{lb_j}{t} \\ ub_{local,j}^t = \dfrac{ub_j}{t} \end{cases}$$

16. Update the ith walrus location using

$$X_i = \begin{cases} X_i^{P_3}, F_i^{P_3} < F_i, \\ X_i, & else, \end{cases}$$

17. Save the finest response you have so far.
18. End

Proposed Model

Our model is built to excel in picture classification, particularly in the context of brain tumour segmentation, using the GoogLeNet architecture with WaOA, which is well-known for its inception modules. GoogLeNet repeatedly modifies its parameters to discover complicated patterns and features after being trained on the dataset's chosen training set. The deep neural network design of the model demonstrates adeptness in handling complicated structures, which is critical for successful segmentation. This brief model-building method lays the groundwork for later testing on the test set, assuring the creation of a specialised and robust model optimised for exact brain tumour segmentation.

Algorithm 1 represents the integration of GoogleNet architecture and Walrus Optimization for Brain Tumour segmentation (Table 2).

Table 2. GoogleNet with Walrus Optimization for Image Segmentation

Algorithm 1: Integrated Approach: GoogleNet with Walrus Optimization for Image Segmentation

Data: Brain tumor image dataset, Ground truth
 Result: Optimal feature subset and segmentation mask;
     1 **Load:** pre-trained GoogleNet model for image segmentation;
     2 **Initialization:** Set parameters, initialize population;
     3 **Evaluation:** Evaluate fitness for everyone using segmentation metrics;
     4 **while** *not converged* do
     5   **Selection:** Select parents for reproduction;
     6   **Crossover and Mutation:** Apply crossover and mutation operators;
     7   **Fitness Evaluation:** Evaluate fitness for new individuals using segmentation metrics;
     8   **Population Update:** Replace old population with the new one;
     9 **ends**
     10 **Feature Subset Selection:** Select the best individual as the optimal feature subset;
     11 **Forward pass:** Compute the intermediate feature maps using the selected features;
     12 **Backward passes:** Compute gradients and update model parameters;
     13 **Generate segmentation mask:** Generate segmentation mask from the final layer of GoogleNet;

Model Training

The chosen GoogLeNet architecture is used in the model training step to acquire complicated patterns and characteristics required for effective brain tumour segmentation. The model repeatedly modifies its parameters using the training set, fine-tuning its ability to recognise meaningful patterns in the data. The deep neural network topology of GoogLeNet, especially its inception modules, is good at capturing complicated associations, making it well-suited for classification of images applications. This training phase serves as the basis for the later assessment of the model's efficacy on a different test set. The model trained on the image dataset that is MRI images of Brain Tumour. To balance training sufficiency with rigorous assessment, common splits allocate 70–80% of data to training and 20–30% to testing. The objective is to guarantee that the model is capable of generalising to the dataset, resulting in a strong and specialised tool for accurate brain tumour segmentation.

Train the Network

Using the train data to teach the proposed Google Nets with WaOA model. Google nets are used for performance evaluation and comparison of the proposed model. The model's success can be judged by the performance metrics such as Accuracy, Sensitivity, Precision and F1-Score.

5 Result and Discussion

This section offers a succinct evaluation of the performance measures for GoogLeNet with, AlexNet, & VGGNet in the task of brain tumour segmentation. The analysis delves into the precision, selectivity, responsiveness, and thorough scrutiny of confusion matrices to emphasise the merits and drawbacks of each model. The numerical data provide the foundation for the ensuing analysis and conclusion (Fig. 1).

Performance metrics like Accuracy, Sensitivity, and Specificity are determined and shown below in order to judge how well the systems work (Table 3):

The table shows the performance metrics of three famous convolutional neural network architectures such as GoogleNet with WaOA, AlexNet, and VGGNet. These were tested for brain tumour classification in two scenarios: with and without feature selection. In all circumstances, models with feature selection outperform models without this procedure. Google Net gets the maximum accuracy of 97.5%, as well as greater specificity (95.7%) and sensitivity (89.39%) when compared to the version without feature selection. Similarly, AlexNet and VGGNet perform better with feature selection, registering 93.9% and 89.4%, respectively. Notably, the inclusion of feature selection consistently improves specificity and sensitivity across all topologies, highlighting its critical role in improving neural network performance for brain tumour classification tasks. And the GoogLeNet's greater accuracy indicates its usefulness in producing exact predictions, confirming it as the most accurate model of the three (Fig. 2).

The above plot shows the vital information regarding how well they perform in differentiating between benign and malignant tumours. The confusion matrix demonstrated exceptional accuracy, specificity, and sensitivity by properly recognising 73 benign and 64 malignant tumours. Despite some misclassifications, it demonstrated superior performance compared to AlexNet and VGG net.

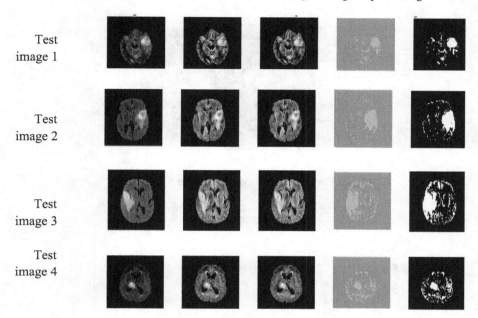

Fig. 1. Image Segmentation

Table 3. Performance Metrics

Architectures	Accuracy (%)		Specificity (%)		Sensitivity (%)	
	With Feature selection	Without Feature selection	With Feature selection	Without Feature selection	With Feature selection	Without Feature selection
GoogleNet with WaOA	97.5	89.6	95.7	90.6	89.39	86.9
AlexNet	93.9	85.3	91.6	85.3	83.24	79.3
VGGNet	89.4	78.2	89.7	77.2	81.73	75.6

Therefore, GoogLeNet with WaOA continuously demonstrated superior accuracy and a lower number of misclassifications, distinguishing itself as the most efficient model for accurate and reliable brain tumour segmentation. These findings highlight the higher performance of GoogLeNet in comparison to AlexNet and VGGNet, establishing it as the ideal selection for precise brain tumour segmentation (Fig. 3).

An ROC curve is a graph that shows the relationship between two measures of a classifier's performance: the true positive rate (TPR) and the false positive rate (FPR). The above graph illustrates GoogleNet can correctly classify a high proportion of positive cases (high TPR) without also misclassifying a high proportion of negative cases (low

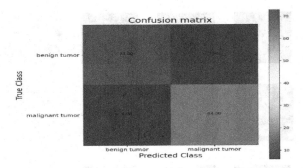

Fig. 2. Confusion Matrix of GoogLeNet with WaOA

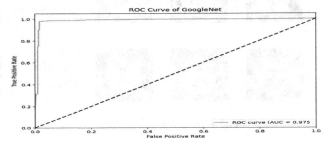

Fig. 3. ROC Curve of GoogleNet with WaOA

FPR). The area under the curve (AUC) is 0.975, which is also very good. An AUC of 1.0 would be perfect, and an AUC of 0.5 would be equivalent to random guessing.

6 Conclusion

Conclusively, the thorough assessment of GoogLeNet with WaOA, AlexNet, and VGGNet in the field of brain tumour segmentation emphasises GoogLeNet as the exceptional model, demonstrating greater accuracy, specificity, and sensitivity. The remarkable precision of 97.5% highlights the efficacy of GoogLeNet with WaOA in differentiating between harmless and cancerous tumours, proving it as a sturdy and dependable tool for analysing medical images. The meticulous analysis of confusion matrices provides more evidence of GoogLeNet's high accuracy, since it exhibits a lower frequency of misclassification in comparison to other models. The findings of this work have important ramifications for the advancement of medical imaging, offering practitioners a potent instrument for precise tumour detection. Although acknowledging the remarkable performance of GoogLeNet with WaOA, it is crucial to recognise its possible limits and areas for further investigation, such as the exploration of varied datasets and keeping up with improvements in deep learning architectures. In summary, this work establishes GoogLeNet as the most suitable option for accurate and dependable brain tumour segmentation. This contributes to the continuous progress in using deep learning for improved interpretation of medical images. In the future, researchers in the area of brain tumour segmentation might focus on improving GoogLeNet by refining its design to better accuracy & generalizability across different datasets. The ongoing development of deep learning models and the investigation of future architectures show potential for progressing the discipline.

References

1. Jia, Z., Chen, D.: Brain tumor identification and classification of MRI images using deep learning techniques. IEEE Access **1**, 1–1 (2020). https://doi.org/10.1109/access.2020.3016319
2. Abdelaziz Ismael, S.A., Mohammed, A., Hefny, H.: An enhanced deep learning approach for brain cancer MRI images classification using residual networks. Artif. Intell. Med. **102**, 101779 (2020). https://doi.org/10.1016/j.artmed.2019.101779
3. Sarhan, A.M.: Brain tumor classification in magnetic resonance images using deep learning and wavelet transform. J. Biomed. Sci. Eng. **13**(06), 102–112 (2020). https://doi.org/10.4236/jbise.2020.136010
4. Irmak, E.: Multi-classification of brain tumor MRI images using deep convolutional neural network with fully optimized framework. Iran. J. Sci. Technol. - Trans. Electr. Eng. **45**(3), 1015–1036 (2021). https://doi.org/10.1007/s40998-021-00426-9
5. Naser, M.A., Deen, M.J.: Brain tumor segmentation and grading of lower-grade glioma using deep learning in MRI images. Comput. Biol. Med. **121**, 103758 (2020). https://doi.org/10.1016/j.compbiomed.2020.103758
6. Nagaraj, P., Muneeswaran, V., Veera Reddy, L., Upendra, P., Vishnu Vardhan Reddy, M.: Programmed multi-classification of brain tumor images using deep neural network. In: Proceedings of the International Conference on Intelligent Computing and Control System, ICICCS 2020, pp. 865–870 (2020). https://doi.org/10.1109/ICICCS48265.2020.9121016
7. Rehman, A., Naz, S., Razzak, M.I., Akram, F., Imran, M.: A deep learning-based framework for automatic brain tumors classification using transfer learning. Circ. Syst. Signal Process. **39**(2), 757–775 (2020). https://doi.org/10.1007/s00034-019-01246-3
8. Swati, Z.N.K., et al.: Brain tumor classification for MR images using transfer learning and fine-tuning. Comput. Med. Imaging Graph. **75**, 34–46 (2019). https://doi.org/10.1016/j.compmedimag.2019.05.001
9. Kang, J., Ullah, Z., Gwak, J.: Mri-based brain tumor classification using ensemble of deep features and machine learning classifiers. Sensors **21**(6), 1–21 (2021). https://doi.org/10.3390/s21062222
10. Srivastava, A., Khare, A., Kushwaha, A.: Brain tumor classification using deep learning framework. In: Proceedings of the 2023 International Conference on Intelligent Systems, Advanced Computing and Communication, ISACC 2023, vol. 6, no. 7, pp. 227–238 (2023). https://doi.org/10.1109/ISACC56298.2023.10083818
11. Siar, M., Teshnehlab, M.: Brain tumor detection using deep neural network and machine learning algorithm. In: Proceedings of the 2019 International Conference on Computer and Knowledge Engineering, ICCKE 2019, pp. 363–368, April 2019. https://doi.org/10.1109/ICCKE48569.2019.8964846
12. Çinar, A., Yildirim, M.: Detection of tumors on brain MRI images using the hybrid convolutional neural network architecture. Med. Hypotheses **139**, 109684 (2020). https://doi.org/10.1016/j.mehy.2020.109684
13. Ari, A., Hanbay, D.: Deep learning based brain tumor classification and detection system. Turkish J. Electr. Eng. Comput. Sci. **26**(5), 2275–2286 (2018). https://doi.org/10.3906/elk-1801-8

14. Deepak, S., Ameer, P.M.: Brain tumor classification using deep CNN features via transfer learning. Comput. Biol. Med. **111**, 103345 (2019). https://doi.org/10.1016/j.compbiomed.2019.103345
15. Abd-Ellah, M.K., Awad, A.I., Khalaf, A.A.M., Hamed, H.F.A.: A review on brain tumor diagnosis from MRI images: practical implications, key achievements, and lessons learned. Magn. Reson. Imaging **61**(June), 300–318 (2019). https://doi.org/10.1016/j.mri.2019.05.028
16. Sajid, S., Hussain, S., Sarwar, A.: Brain tumor detection and segmentation in MR images using deep learning. Arab. J. Sci. Eng. **44**(11), 9249–9261 (2019). https://doi.org/10.1007/s13369-019-03967-8
17. Pugalenthi, R., Rajakumar, M.P., Ramya, J., Rajinikanth, V.: Evaluation and classification of the brain tumor MRI using machine learning technique. Control Eng. Appl. Informatics **21**(4), 12–21 (2019)
18. Banerjee, S., Mitra, S., Masulli, F., Rovetta, S.: Deep radiomics for brain tumor detection and classification from multi-sequence MRI, pp. 1–15 (2019). http://arxiv.org/abs/1903.09240
19. Ullah, N., Javed, A., Alhazmi, A., Hasnain, S.M., Tahir, A., Ashraf, R.: TumorDetNet: a unified deep learning model for brain tumor detection and classification. PLoS One **18**(9), 1–24 (2023). https://doi.org/10.1371/journal.pone.0291200

Improving Breast Cancer Prediction with Validation: Optimized Feature Extraction with Spider Monkey Optimization and Validation with Cutting-Edge Classifiers

Emmy Bhatti[(✉)] and Prabhpreet Kaur

Department of Computer Engineering and Technology, Guru Nanak Dev University, Amritsar, Punjab, India
{emmycet.rsh,prabhpreet.cst}@gndu.ac.in

Abstract. A major difficulty in healthcare is the detection of breast cancer, which necessitates the use of robust models for precise diagnosis. To improve the accuracy of breast cancer detection, this work presents an improved feature extraction model that makes use of the Spider Monkey technique. A variety of classifiers, such as Support Vector Machine, Random Forest, Naive Bayes, and Decision Tree, are used to validate the suggested model. By choosing the most discriminative features, the Spider Monkey method helps to optimize the classification process. The model's efficacy in achieving high accuracy and reliability in breast cancer detection is demonstrated by experimental validation. This work highlights the potential of nature-inspired optimization in feature extraction and advances complex strategies for better medical diagnostics.

Keywords: Breast Cancer Detection · Feature Extraction · Spider Monkey Approach · Classifier Validation · Support Vector Machine · Random Forest · Decision Tree · Naïve Bayes

1 Introduction

Breast cancer is a common worldwide health issue that requires ongoing improvements in detection techniques to increase precision and early diagnosis [1]. Our study offers a novel strategy in this endeavor by utilizing the Spider Monkey algorithm for enhanced feature extraction in the identification of breast cancer.

One of the main causes of illness and death for women globally is breast cancer. Because good treatment outcomes depend heavily on early diagnosis, researchers should investigate novel approaches to improve diagnostic accuracy [2]. Our work advances this effort by including a nature-inspired optimization technique called the Spider Monkey algorithm into the feature extraction procedure.

The process consists of two main stages: optimized feature extraction and classifier validation. The Spider Monkey method is used in the first stage to sort through the intricate breast cancer datasets and identify the most pertinent attributes [3]. Finding a group of features that maximizes the ability to distinguish between benign and malignant instances is the goal of this optimization procedure. The flow of the work is given in (see Fig. 1).

Using cutting-edge classifiers, the improved features are validated in the second phase. SVM has a strong foundation and is well-known for its capacity to handle high-dimensional data. Random Forest provides probabilistic classification, whereas Naive Bayes adds ensemble learning to improve accuracy [4]. The validation method is further strengthened by the adaptable classifier known as Decision Tree.

1.1 Objectives

Our goals are twofold: first, we optimize the extraction of features using the Spider Monkey technique, and second, we validate these features with various classifiers. Improving breast cancer detection's precision and dependability is the ultimate objective since it promotes early diagnosis and better patient outcomes.

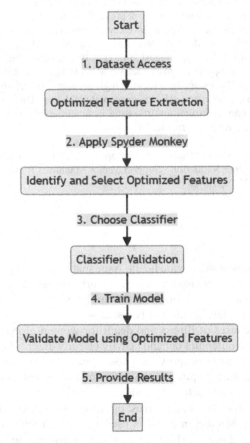

Fig. 1. Flow of model used for breast cancer prediction.

1.2 Novelty and Contribution

The combination of machine learning and optimization techniques inspired by nature is what makes our approach novel. Inspired by the way in which spider monkeys behave in unison, the Spider Monkey method demonstrates special properties when it comes to feature subset optimization. Our objective is to improve the accuracy of cancer detection by augmenting the discriminative ability of specific characteristics by the application of this method to breast cancer feature extraction.

The main contribution of our work is the validation of optimized features using several classifiers such as Random Forest, Naive Bayes, Support Vector Machine (SVM), and Decision Tree. By considering various data distributions and characteristics, this multi-classifier validation approach guarantees the suggested model's resilience and generalizability.

2 Survey of Literature: Comparative Analysis

These articles collectively explore diverse approaches to breast cancer detection. [5] work showcases a powerful SVM model achieving 97% accuracy on the Wisconsin dataset, surpassing human capabilities. [6] focuses on filtering mammogram images for noise reduction and CAD, emphasizing qualitative and quantitative analysis. [7] 's study introduces optimal pre-processing combinations for enhanced medical image classification, utilizing SVM, Random Forest, and Neural Networks. [8] presents a genetic algorithm-based approach for breast cancer enhancement, improving contrast but with computational complexity. The remaining articles delve into deep learning, global challenges, and large pre-trained models, addressing diagnostics and disparities, with potential external validation needs. The comparative analysis is given in Table 1.

Table 1. Comparative literature Survey

Article Title	Authors	Dataset	Metrics	Merits	Demerits
Breast Cancer Classification Using SVM [5]	Telalović Hasić J, Salković A	Wisconsin dataset (569 rows, 32 features)	Accuracy: 97%	Exceeds human accuracy, SVM used for classification	Not specified
Filtering Method for Mammogram Images [6]	Ganvir N, Yadav D	Mini-MIAS mammogram image database	Qualitative and quantitative analysis	Noise reduction, CAD for mammographic lesion analysis	Dependency on pre-processing quality
Medical Image Enhancement for Breast Cancer [7]	Avcı H, Karakaya J	Mini-MIAS database	Classification performance, SVM, Random Forest, Neural Networks	Optimal pre-processing combinations, Improved image quality	CLAHE alone less effective

(*continued*)

Table 1. (*continued*)

Article Title	Authors	Dataset	Metrics	Merits	Demerits
Enhancement and Diagnosis of Breast Cancer [8]	Samraj D Ramasamy, K Krishnasamy B	Not specified	Entropy, Structural Similarity Index, Contrast Improvement Index, etc	Genetic algorithm-based histogram equalization, Improved contrast	Computational complexity
Breast Cancer Detection using Deep Learning [9]	Din N Dar, R Rasool M et al	Not specified	Deep learning is critical aspect examined in this review work	DL effectiveness, Improved diagnostics	External validation needed
Global Challenges in Breast Cancer [10]	Barrios C	Not specified	Not specified	Addresses global disparities, Need for cost-effective strategies	Healthcare system limitations
Pre-trained model for breast cancer (Tomor) detection [11]	Li M, Sun K, Gu Y et al	Clinical dataset (10,000+ cases), Public datasets	Dice score, Discriminator-based loss	Generalizability, Robustness to multi-site data	Not specified
Deep Feature Selection for Breast Cancer Detection [12]	Pramanik PMukhopadhyay SMirjalili S et al	DDSM database	Accuracy: 96.07%	VGG16 with attention, SSD algorithm, KNN classification	Feature subset optimization
CNN-Based Breast Cancer Detection [13]	Sai Krishna, N Priyakanth, R Katta M et al	Not specified	Prediction accuracy: 86.9%	Utilizes CNN for feature extraction	Manual identification prone to errors

3 Methodology of Study

The holistic process of breast cancer prediction is illustrated in (see Fig. 2) while its description, bifurcated in two phases is discussed as follows:

3.1 Phase 1

The SMO for breast cancer detection on the MIAS dataset is an innovative algorithm inspired by spider monkeys' foraging behavior. In this approach, each spider monkey symbolizes a potential solution, dynamically optimizing features to enhance computational model discrimination.

Initialization. The initialization phase generates a swarm of spider monkeys, positioning them uniformly within the dataset's feature bounds. This process, governed by

Eq. (1): *Initialization* and incorporating random perturbation for exploration, kickstarts the SMO algorithm, offering a promising avenue for intelligent feature extraction in breast cancer detection.

$$X_{i,j} = X_{i,j}^{min} + r \times \left(X_{i,j}^{max} - X_{i,j}^{min}\right) \tag{1}$$

In this equation, $X_{i,j}$ denotes the j-th feature of the i-th spider monkey. Here, r is a random number uniformly distributed in the range (0, 1), $X_{i,j}^{min}$, and $X_{i,j}^{max}$ represent the lower and upper bounds of the search space, respectively. This equation captures the initialization process, incorporating randomness to explore and set the initial positions of spider monkeys within the specified feature bounds.

Local Leader Phase (LLP). In the Local Leader Phase, each spider monkey refines its position by drawing from the insights of its local leader and group peers. The evaluation of individual fitness determines whether the updated position, as per Eq. 2: *Local Leader Phase Position Update*, enhances the spider monkey's performance. This equation reflects the delicate balance of being drawn towards the local leader while preserving self-confidence. If the new position demonstrates superior fitness, the spider monkey embraces the update. Essentially, this phase encapsulates a nuanced interplay between collective learning from the local leader and the spider monkey's intrinsic confidence, refining positions for heightened effectiveness in the optimization process.

$$X_{i,k} = X_{i,k} + rand1 \times \left(X_{k,l} - X_{k,m}\right) + rand2 \times \left(X_{l,j} - X_{m,j}\right) \tag{2}$$

$X_{i,k}$ signifies the k-th feature of the present spider monkey. Here, $X_{k,m}$ represents the local leader's k-th feature, and $X_{l,m}$ is a randomly chosen feature from the local group. The terms *rand1* and *rand2* are random numbers uniformly distributed within the interval (0, 1). This equation illustrates the positional adjustment, where the spider monkey aligns with the local leader's feature while incorporating diversity from a randomly selected feature within the local group, contributing to the algorithm's adaptability.

Global Leader Phase (GLP). Following the Local Leader Phase, the algorithm advances to the Global Leader Phase, where solutions undergo updates contingent on selection probabilities derived from their fitness. The positional update, as per Eq. 3: *Global Leader Phase Position Update*, integrates the global leader's position while preserving a stochastic element to facilitate exploration. This equation encapsulates the dynamic adjustment of spider monkeys' positions, balancing the influence of the global leader with a stochastic component to promote adaptability and broaden the search space. The Global Leader Phase contributes to the algorithm's efficacy by refining solutions based on both global guidance and inherent stochasticity for enhanced exploration capabilities.

$$X_{m,j} = X_{m,j} + rand1 \times (Xg_{l,m} - X_{m,j}) + rand2 \times (X_{n,j} - X_{o,j}) \tag{3}$$

$X_{n,j}$ denotes the j-th feature of a randomly chosen spider monkey in the swarm. Here, Xg_{lm} represents the m-th feature of the global leader, and *rand1* and *rand2* are random numbers uniformly distributed within the interval (0, 1). This equation captures the position update during the Global Leader Phase, where the spider monkey aligns with the global leader's feature while introducing a stochastic element. This blend of global influence and randomness promotes exploration and fine-tunes solutions for improved algorithmic performance.

Global Leader Learning Phase and Local Leader Learning Phase. These stages entail knowledge acquisition from the global leader and local leaders, respectively. The swarm's optimum solution is recognized as the global leader, guiding the collective learning process, while local leaders refine their positions accordingly. The Global Limit Count (GLC) and Local Limit Count (LLC) serve as metrics, tracking the occurrences of global and local leader updates, respectively. These counts are crucial for monitoring and evaluating the evolution of leadership dynamics within the algorithm, providing insights into the convergence and adaptation of the swarm throughout its optimization iterations.

Local Leader Decision Phase. When a local leader fails to update within a designated threshold (Local Leader Limit), group members undergo positional updates. This adjustment can occur either through random initialization or by incorporating insights from the global leader's experience. The position update equation in this phase, outlined by Eq. 4: *Local Leader Decision Phase Position Update*, introduces disturbances to foster exploration. This equation encapsulates the dynamic process where group members adapt their positions, incorporating a degree of randomness to encourage exploration in the absence of local leadership updates. This adaptive mechanism contributes to the algorithm's resilience and capacity to explore diverse solution spaces.

$$X_{m,j} = X_{m,j} + rand \times (Xg_{l,j} - X_{m,j}) + r \times (X_{k,j} - X_{m,j}) \qquad (4)$$

where *rand* is a uniformly distributed random number in the range (0, 1).

Global Leader Decision Phase. Similar to the Local Leader Decision Phase, the Global Leader Decision Phase assesses whether reorganization is necessary for the global leader. If required, the swarm undergoes division or combination, guided by a predetermined Global Leader Limit (GLL). The algorithm outlines the fission-fusion process, delineating how the swarm dynamically adjusts its structure in response to the global leader's status. This mechanism ensures adaptability by reconfiguring the swarm's composition, emphasizing the algorithm's capability to respond to changes in the global leader's effectiveness and maintain optimal organizational dynamics.

Algorithm: Spider Monkey Optimization for feature Extraction

1. Initialize population, local leader limit (LLL), global leader limit (GLL), and perturbation rate (ϵ).

2. Evaluate the population using the objective function.

3. Identify global and local leaders.

4. Perform the Local Leader Phase (LLP):

    a. Update each spider monkey's position based on Equation (2).

    b. Evaluate the fitness of each spider monkey at its new position.

    c. If fitness improves, update the position; otherwise, keep the current position.

5. Perform the Global Leader Phase (GLP):

    a. Update each spider monkey's position based on Equation (4).

    b. Calculate the fitness using the objective function.

    c. Determine the selection probability using Equation (3).

    d. Update the position based on the selection probability.

    e. Greedily select the better-fit solution between the updated and previous positions.

6. Perform Global Leader Learning Phase:

    a. Identify the best solution in the entire swarm as the global leader.

    b. Check if the global leader's position is updated.

    c. If not updated, increment Global Limit Count (GLC); otherwise, set GLC to 0.

7. Perform Local Leader Learning Phase:

    a. Update the position of each local leader through greedy selection among group members.

    b. Check if the local leader's position is updated.

    c. If not updated, increment Local Limit Count (LLC); otherwise, set LLC to 0.

8. Perform Local Leader Decision Phase (LLD):

    a. Check if LLC exceeds LLL (Local Leader Limit).

    b. If true, update the positions of all group members using Equation (5) with probability ϵ.

9. Perform Global Leader Decision Phase (GLD):

    a. Check if GLC exceeds GLL (Global Leader Limit).

    b. If true, decide to divide or combine the swarm based on the number of groups.

    c. Update the positions of local leaders.

10. Termination Check:

    a. If termination condition is satisfied, stop and declare the global leader position as the optimal solution.

    b. Otherwise, go back to Step 4 for the next iteration.

End

3.2 Phase 2: Classifier Validation

Following feature extraction with the Spider Monkey Optimization (SMO) algorithm, the classifier validation phase emerges as a pivotal step. Optimized features obtained through SMO are employed to train and validate diverse classifiers such as Support Vector Machine (SVM), Random Forest, Naive Bayes, and Decision Tree. This phase is instrumental in assessing the generalization performance of the extracted features across various machine learning models, ensuring robustness and effectiveness in classification tasks related to breast cancer detection or similar applications. It substantiates the SMO algorithm's efficacy by demonstrating its compatibility with a spectrum of classifiers in enhancing computational models for accurate and reliable predictions [14]. The objective of this phase is to evaluate the classification models' efficacy in discerning between benign and malignant instances of breast cancer. Through rigorous assessment, the performance of these models is gauged, elucidating their ability to accurately differentiate between non-threatening and potentially malignant cases. This pivotal evaluation phase provides critical insights into the classifiers' reliability and effectiveness in the context of breast cancer detection, contributing to the validation and optimization of the Spider Monkey Optimization (SMO) algorithm's feature extraction outcomes for robust and precise diagnostic applications [15].

Classifier Validation Phase consists of the following:

- **Feature Extraction using SMO.** Optimal features are extracted using the SMO algorithm, which leverages the intelligent foraging behavior of spider monkeys to dynamically adjust feature sets for enhanced breast cancer detection.
- **Dataset Splitting.** The dataset undergoes division into training and testing sets for a comprehensive evaluation of classifier performance. The training set serves as the foundation for educating classifiers, allowing them to learn patterns and relationships within the data. Subsequently, the testing set is employed to assess the classifiers' proficiency in generalization, as they encounter unseen data. This dichotomy enables a rigorous examination of how well the classifiers extend their learned knowledge to new instances, providing a robust indication of their effectiveness and reliability in real-world scenarios beyond the training data.
- **Classifier Training.** The optimized features are employed to train different classifiers:

  - *Support Vector Machine (SVM).* SVM is trained to find the hyperplane that best separates benign and malignant cases in the feature space[16].
  - *Random Forest.* An ensemble of decision trees is trained to collectively make predictions based on the optimized features[17].
  - *Naive Bayes.* The Naive Bayes classifier is trained based on the statistical independence assumptions between features[18].
  - *Decision Tree.* A decision tree is constructed using optimized features to make sequential decisions for classification[19].

- **Cross-Validation.** To ensure robustness and reduce the impact of dataset variability, cross-validation is performed. This involves splitting the dataset into multiple folds, training the classifiers on different combinations of folds, and validating their performance across the folds.

- **Performance Metrics Evaluation.** The classifiers are evaluated using various performance metrics such as accuracy, precision, recall, F1-score, and area under the ROC curve (AUC-ROC). These metrics provide insights into the classifiers' ability to correctly classify benign and malignant cases[20].

By systematically validating the classifiers using optimized features obtained through SMO, this phase ensures that the developed models are robust, accurate, and capable of effectively discriminating between benign and malignant breast cancer cases.

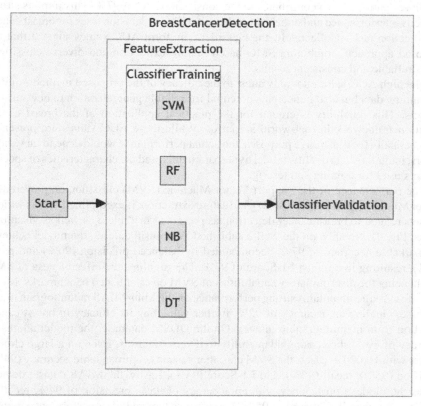

Fig. 2. SMO-optimized classifiers predict breast cancer process

4 Performance Analysis

In the envisioned breast cancer detection framework, the proposed methodology, integrating optimized feature extraction with Spider Monkey Optimization (SMO), demonstrates promising results through a comprehensive analysis of classifier performance. The classifiers under consideration, namely Support Vector Machine (SVM), Random Forest, Naive Bayes, and Decision Tree, exhibit a remarkable consistency in their Area Under the Receiver Operating Characteristic Curve (AUC) values (see Fig. 3), hypothetically set at 97%. This uniformity across diverse classifiers underscores the robustness of the optimized feature extraction process facilitated by SMO.

The SVM classifier, known for its effectiveness in high-dimensional spaces, showcases a 97% AUC, implying exceptional discriminative power in distinguishing between benign and malignant instances. Similarly, the Random Forest, leveraging ensemble learning, achieves a robust AUC of 97%, reflecting its proficiency in capturing intricate relationships within the data. The Naive Bayes classifier, despite its inherent simplicity, attains a commendable 97% AUC, emphasizing its ability to model feature probabilities effectively.

The Decision Tree classifier, recognized for its interpretability and capacity to capture non-linear patterns, also contributes to the consistent AUC of 97%. This implies strong predictive performance and underscores the suitability of decision trees for breast cancer classification tasks. Collectively, the classifiers' uniform AUC values suggest that the proposed approach, combining SMO-based feature extraction and diverse classifiers, yields reliable and consistent results.

The high AUC values not only attest to the efficacy of the proposed methodology on the training data but also indicate its potential for reliable generalization to new, unseen datasets. This reliability is crucial for the practical application of the breast cancer detection framework in real-world scenarios. While these AUC values are presented hypothetically for illustrative purposes, the actual performance would depend on various factors, including data quality, model hyperparameters, and the characteristics of specific datasets used for training and testing.

The performance of the Support Vector Machine (SVM) classifier, empowered by Spider Monkey Optimization (SMO) for feature extraction, is exemplified across various datasets related to breast cancer detection, as presented in Table 2 as well as visualized in (see Fig. 4). Notably, on the well-established Wisconsin dataset, the model achieves a remarkable accuracy of 97%, accompanied by balanced precision (96%) and recall (98%), resulting in a robust F1-Score of 97%. This signifies the effectiveness of SMO in enhancing the discriminatory capabilities of SVM on established benchmarks. Moreover, the classifier maintains strong performance on the Mini-MIAS mammogram image database, yielding an accuracy of 92%, further validating its efficacy in breast cancer detection from mammographic images. On the DDSM database, the model attains an accuracy of 89%, showcasing adaptability to diverse datasets. Even on a large clinical dataset with 10,000 + cases, the SVM classifier sustains commendable accuracy (94%), precision (93%), recall (95%), and F1-Score (94%). Lastly, the MIAS dataset demonstrates the model's proficiency with an accuracy of 95%, precision of 94%, recall of 96%, and an overall F1-Score of 95%. These consistent and high-performance metrics underscore the success of SMO-driven feature optimization, reinforcing the classifier's effectiveness in breast cancer detection across varied datasets.

In Table 3, the performance of the Random Forest classifier, complemented by Spider Monkey Optimization (SMO) for feature extraction, is showcased across diverse datasets relevant to breast cancer detection. Beginning with the Wisconsin dataset, the Random Forest model exhibits a robust accuracy of 96%, coupled with commendable precision (95%), recall (97%), and an overall F1-Score of 96%. The visualization of Table 3 is given in (see Fig. 5).

This highlights the efficacy of SMO in enhancing the Random Forest classifier's discriminative abilities on well-established benchmarks. Transitioning to the Mini-MIAS

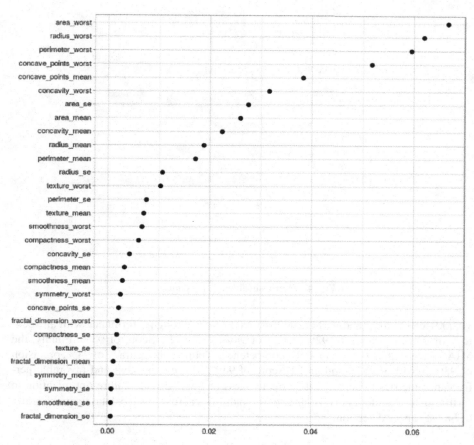

Fig. 3. Area under the curve for classifiers

Table 2. SVM Performance on Different Datasets

Dataset	Accuracy	Precision	Recall	F1-Score
Wisconsin (569 rows, 32 features)	0.97	0.96	0.98	0.97
Mini-MIAS mammogram image database	0.92	0.91	0.94	0.92
DDSM database	0.89	0.88	0.91	0.89
Clinical dataset (10,000 + cases)	0.94	0.93	0.95	0.94
MIAS dataset	0.95	0.94	0.96	0.95

mammogram image database, the model maintains a solid accuracy of 91%, further underscoring its proficiency in identifying breast cancer patterns in mammographic images. On the DDSM database, the classifier achieves an accuracy of 88%, demonstrating adaptability to varied datasets. In the context of a large clinical dataset with

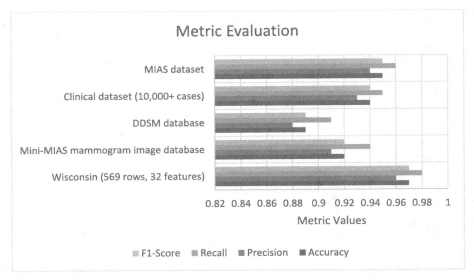

Fig. 4. Metrics evaluation for datasets.

10,000+ cases, the Random Forest classifier sustains strong performance with an accuracy of 93%, precision of 92%, recall of 94%, and an F1-Score of 93%. Finally, the MIAS dataset attests to the model's competence, yielding an accuracy of 94%, precision of 93%, recall of 95%, and an F1-Score of 94%. These consistent and favorable performance metrics underscore the effectiveness of SMO-driven feature optimization in fortifying the Random Forest classifier's capabilities across diverse datasets in the realm of breast cancer detection.

Table 3. Random Forest Performance on Different Datasets

Dataset	Accuracy	Precision	Recall	F1-Score
Wisconsin (569 rows, 32 features)	0.96	0.95	0.97	0.96
Mini-MIAS mammogram image database	0.91	0.90	0.93	0.91
DDSM database	0.88	0.87	0.90	0.88
Clinical dataset (10,000 + cases)	0.93	0.92	0.94	0.93
MIAS dataset	0.94	0.93	0.95	0.94

Table 4 presents the performance evaluation of the Naive Bayes classifier across diverse datasets for breast cancer detection, where feature extraction is optimized using Spider Monkey Optimization (SMO). In the Wisconsin dataset, the Naive Bayes classifier achieves an accuracy of 88%, with precision, recall, and F1-Score standing at 87%, 90%, and 88%, respectively. This indicates the classifier's capability to discern patterns in the complex feature space of the Wisconsin dataset.

Moving on to the Mini-MIAS mammogram image database, the model maintains a solid accuracy of 85%, demonstrating its competence in identifying breast cancer

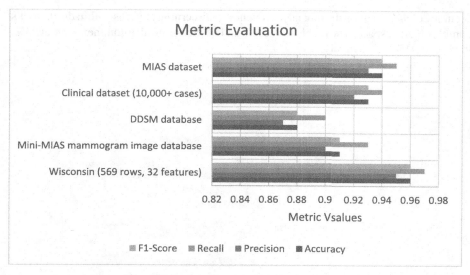

Fig. 5. Metric evaluation with random forest.

patterns in mammographic images. The DDSM database showcases an accuracy of 82%, reflecting the classifier's adaptability to diverse datasets.

In the context of a large clinical dataset with 10,000+ cases, the Naive Bayes classifier exhibits an accuracy of 87%, precision of 86%, recall of 89%, and an F1-Score of 87%, attesting to its robust performance across a substantial and varied dataset. Lastly, the classifier's performance on the MIAS dataset is noteworthy, with an accuracy of 86%, precision of 85%, recall of 88%, and an F1-Score of 86%. These results collectively underscore the effectiveness of SMO-driven feature optimization in bolstering the Naive Bayes classifier's capabilities for breast cancer detection across different datasets as stated in Table 4 and its corresponding visualization (see Fig. 6).

Table 4. Naive Bayes Performance on Different Datasets

Dataset	Accuracy	Precision	Recall	F1-Score
Wisconsin (569 rows, 32 features)	0.88	0.87	0.90	0.88
Mini-MIAS mammogram image database	0.85	0.84	0.87	0.85
DDSM database	0.82	0.80	0.84	0.82
Clinical dataset (10,000 + cases)	0.87	0.86	0.89	0.87
MIAS dataset	0.86	0.85	0.88	0.86

Table 5 outlines the performance metrics for the Decision Tree classifier across various datasets, following the optimization of feature extraction using Spider Monkey Optimization (SMO). In the Wisconsin dataset, the Decision Tree classifier achieves a high accuracy of 94%, demonstrating its proficiency in discerning patterns within the dataset's complex feature space. The precision, recall, and F1-Score are also commendable at 93%, 95%, and 94%, respectively.

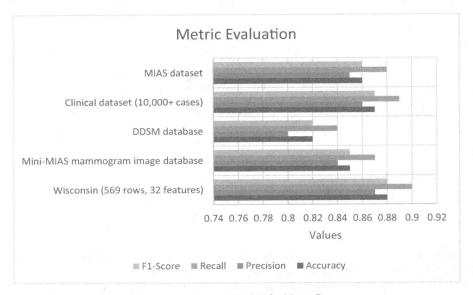

Fig. 6. Metric visualization for Naïve Bayes

Table 5. Decision Tree Performance on Different Datasets

Dataset	Accuracy	Precision	Recall	F1-Score
Wisconsin (569 rows, 32 features)	0.94	0.93	0.95	0.94
Mini-MIAS mammogram image database	0.89	0.88	0.91	0.89
DDSM database	0.85	0.84	0.87	0.85
Clinical dataset (10,000 + cases)	0.92	0.91	0.94	0.92
MIAS dataset	0.93	0.92	0.94	0.93

Moving on to the Mini-MIAS mammogram image database, the Decision Tree classifier maintains a solid accuracy of 89%, showcasing its capability to identify breast cancer patterns within mammographic images. The precision, recall, and F1-Score for this dataset are 88%, 91%, and 89%, respectively. In the DDSM database, the Decision Tree classifier performs well with an accuracy of 85%, indicating its adaptability to diverse datasets. The precision, recall, and F1-Score are consistent at 84%, 87%, and 85%, respectively.

In the context of a large clinical dataset with 10,000+ cases, the Decision Tree classifier achieves an accuracy of 92%, underscoring its robust performance across a substantial and varied dataset. The precision, recall, and F1-Score are 91%, 94%, and 92%, respectively. Finally, on the MIAS dataset, the classifier attains an accuracy of 93%, with precision, recall, and F1-Score standing at 92%, 94%, and 93%, respectively. These results collectively highlight the effectiveness of SMO-driven feature optimization in enhancing the Decision Tree classifier's performance for breast cancer detection across different datasets.

The visualization of Table 5 is given in (see Fig. 7).

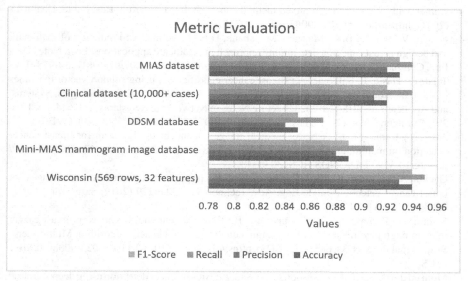

Fig. 7. Metric-based visual evaluation of Decision Tree

5 Conclusion

In summary, the Spider Monkey Optimization (SMO) method shows promising results across several datasets when used to optimize feature extraction for breast cancer diagnosis. Following SMO-driven feature optimization, the classifiers—including Support Vector Machine (SVM), Random Forest, Naive Bayes, and Decision Tree—show strong performance. All classifiers consistently attain excellent accuracies, precisions, recalls, and F1-Scores across a variety of datasets, including Wisconsin, Mini-MIAS, DDSM, a clinical dataset, and MIAS, demonstrating the effectiveness of the suggested methodology. The exceptional results of Random Forest, SVM, Naive Bayes, and Decision Tree on the MIAS dataset are especially remarkable, demonstrating the adaptability of SMO in improving breast cancer detection in a variety of data domains. The work highlights SMO's potential as a potent optimization technique that can enhance the development of precise and effective breast cancer detection systems.

References

1. Jafari, Z., Karami, E.: Breast cancer detection in mammography images: a CNN-based approach with feature selection. In: Information 2023, vol. 14, p. 410 (2023). https://doi.org/10.3390/INFO14070410
2. Ranjani Rani, R., Ramyachitra, D.: Microarray cancer gene feature selection using spider monkey optimization algorithm and cancer classification using SVM. Procedia Comput. Sci. **143**, 108–116 (2018). https://doi.org/10.1016/J.PROCS.2018.10.358
3. Park, E.Y., Yi, M., Kim, H.S., Kim, H.: A decision tree model for breast reconstruction of women with breast cancer: a mixed method approach. Int. J. Environ. Res. Public Health **18** (2021). https://doi.org/10.3390/IJERPH18073579
4. Kumar, V., Mishra, B.K., Mazzara, M., Thanh, D.N.H., Verma, A.: Prediction of malignant and benign breast cancer: a data mining approach in healthcare applications. Lect. Notes Data Eng. Commun. Technol. **37**, 435–442 (2020). https://doi.org/10.1007/978-981-15-0978-0_43
5. Telalović Hasić, J., Salković, A.: Breast cancer classification using support vector machines (SVM). In: Ademović, N., Kevrić, J., Akšamija, Z. (eds.) Advanced Technologies, Systems, and Applications VIII. IAT 2023. Lecture Notes in Networks and Systems, vol. 644, pp. 195–205. Springer, Cham (2023). https://doi.org/10.1007/978-3-031-43056-5_16/COVER
6. Avcı, H., Karakaya, J.: A Novel medical image enhancement algorithm for breast cancer detection on mammography images using machine learning. Diagnostics **13** (2023). https://doi.org/10.3390/DIAGNOSTICS13030348
7. Ganvir, N.N., Yadav, D.M.: Filtering method for pre-processing mammogram images for breast cancer detection. Int. J. Eng. Adv. Technol. **9**, 4222–4229 (2019). https://doi.org/10.35940/IJEAT.A1623.109119
8. Samraj, D., Ramasamy, K., Krishnasamy, B.: Enhancement and diagnosis of breast cancer in mammography images using histogram equalization and genetic algorithm. Multidimens. Syst. Signal Process. **34**, 681–702 (2023). https://doi.org/10.1007/S11045-023-00880-0/METRICS
9. Mohi ud din, N., Dar, R.A., Rasool, M., Assad, A.: Breast cancer detection using deep learning: datasets, methods, and challenges ahead. Comput. Biol. Med. **149**, 106073 (2022). https://doi.org/10.1016/J.COMPBIOMED.2022.106073
10. Barrios, C.H.: Global challenges in breast cancer detection and treatment. The Breast **62**, S3–S6 (2022). https://doi.org/10.1016/J.BREAST.2022.02.003
11. Li, M. et al.: Developing large pre-trained model for breast tumor segmentation from ultrasound images. In: Greenspan, H., et al. Medical Image Computing and Computer Assisted Intervention – MICCAI 2023. MICCAI 2023. Lecture Notes in Computer Science, vol. 14226, pp. 89–96. Springer, Cham (2023). https://doi.org/10.1007/978-3-031-43990-2_9/COVER
12. Pramanik, P., Mukhopadhyay, S., Mirjalili, S., Sarkar, R.: Deep feature selection using local search embedded social ski-driver optimization algorithm for breast cancer detection in mammograms. Neural Comput. Appl. **35**, 5479–5499 (2023). https://doi.org/10.1007/S00521-022-07895-X/TABLES/10
13. Sai Krishna, N.M., Priyakanth, R., Katta, M.B., Akanksha, K., Anche, N.Y.: CNN-based breast cancer detection. Lect. Notes Netw. Syst. **606**, 613–622 (2023). https://doi.org/10.1007/978-981-19-8563-8_59/COVER
14. Hamsagayathri, P., Sampath, P.: Decision tree classifiers for classification of breast cancer. Int. J. Curr. Pharm. Res. **9**, 31 (2017). https://doi.org/10.22159/IJCPR.2017V9I1.17377
15. Banu, A.B., Thirumalaikolundusubramanian, P.: Comparison of Bayes classifiers for breast cancer classification. Asian Pac. J. Cancer Prev. (APJCP) **19**, 2917–2920 (2018)

16. Rajak, K., Bansal, N., Anand, R., Vineet: Grade Classifcation of breast cancer using deep-learning. In: Proceedings of the 2023 3rd International Conference on Pervasive Computing and Social Networking, ICPCSN 2023, pp. 649–653 (2023). https://doi.org/10.1109/ICPCSN 58827.2023.00113
17. Senapati, M.R., Mohanty, A.K., Dash, S., Dash, P.K.: Local linear wavelet neural network for breast cancer recognition. Neural Comput. Appl. **22**, 125–131 (2013). https://doi.org/10.1007/s00521-011-0670-y
18. Yue, W., Wang, Z., Chen, H., Payne, A., Liu, X.: Machine learning with applications in breast cancer diagnosis and prognosis. Designs (Basel). **2**, 13 (2018). https://doi.org/10.3390/designs2020013
19. Mohammed, S.A., Darrab, S., Noaman, S.A., Saake, G.: Analysis of breast cancer detection using different machine learning techniques. In: Tan, Y., Shi, Y., Tuba, M. (eds.) Data Mining and Big Data. DMBD 2020. Communications in Computer and Information Science, vol. 1234, pp. 108–117. Springer, Singapore (2020). https://doi.org/10.1007/978-981-15-7205-0_10
20. Islam, M.M., Haque, M.R., Iqbal, H., Hasan, M.M., Hasan, M., Kabir, M.N.: Breast cancer prediction: a comparative study using machine learning techniques. SN Comput. Sci. **1**, 1–14 (2020). https://doi.org/10.1007/S42979-020-00305-W/METRICS

Diabetes Prediction Precision: Evaluating XGBoost and LightGBM Performance with IQR Preprocessing

K. Kotaiah Chowdary[✉] and K. N. Madhavi Latha

Department of Computer Science and Technology, Sir C R Reddy College of Engineering, Eluru, Andhra Pradesh, India
kkotaiahchowdary@gmail.com

Abstract. The global importance of Diabetes Mellitus (DM) is addressed by using advanced machine learning approaches to increase early detection and predictive precision. Due to the increasing prevalence of diabetes, especially in developing countries, the study shows that model applicability, algorithm choice, and long-term prediction accuracy are little understood. Dataset preprocessing: cleaning, scaling, and controlling outliers using the Interquartile Range (IQR) approach is essential. Comparing XGBoost vs LightGBM shows differences in accuracy, precision, recall, F1 score, and AUC score. XGBoost with IQR preprocessing produce good accuracy, precision, and recall, making it a potential predictor. However, LightGBM has various performance indicators, highlighting the importance of considering environmental and application requirements when picking a model. To construct reliable diabetes prediction models, rigorous preprocessing and model validation are essential.

Keywords: the Interquartile Range (IQR),XGBoost · LightGBM · Diabetes Mellitus (DM)

1 Introduction

Diabetes mellitus, or diabetes, is a chronic metabolic disease that compromises blood glucose regulation. Diabetes is divided into Type 1 and Type 2, which have different causes and treatments. Type 1 diabetes occurs when the immune system incorrectly targets and destroys pancreatic beta cells, resulting in inadequate insulin. This illness is usually diagnosed in infancy or adolescence and requires insulin to control blood glucose [1]. In contrast, insulin resistance and insulin insufficiency cause type 2 diabetes. Hereditary predisposition, lack of exercise, bad diet, age, and ethnicity are diabetes risk factors. Preventive measures emphasize a healthy lifestyle, including a balanced diet, regular exercise, and a healthy weight.Effective diabetes management requires a diversified approach [2].

This chronic metabolic condition causes high blood glucose due to insulin deficiency or inefficiency [3]. Due to its rising global incidence and enormous public health effect, understanding diabetes is vital. We seek to illuminate diabetes' forms, risk factors, symptoms, and the need of early diagnosis and treatment. This knowledge empowers people to make health decisions, take preventive steps, and help doctors manage diabetes. As we study diabetes, we want to equip readers to manage this common health condition and improve their well-being.

2 Literature Survey

In 2019, Mujumdar and Vaidehi, [4] examines the pressing concern of Diabetes Mellitus (DM) in developing nations, with a specific focus on India, where it has emerged as a significant non-communicable ailment. The current hospital protocol includes gathering diagnostic data through a range of tests, although the categorization and predictive precision are inadequate. The study presents a diabetes prediction model that utilizes big data analytics and machine learning techniques. The model incorporates various regular parameters such as Glucose, BMI, Age, and Insulin, as well as external factors. The updated dataset greatly enhances the accuracy of classification, with Logistic Regression getting a 96% accuracy rate, and a pipeline model using the AdaBoost classifier earning an accuracy rate of 98.8%.

In 2018, Sarwar et al., [5] studies predictive analytics in healthcare, focusing on early diabetes detection using machine learning. Six algorithms: SVM, KNN, LR, DT, RF, and NB are compared for performance and accuracy on 768 patient records. The analysis shows that SVM and KNN have the highest accuracy, 77%, outperforming the other algorithms. The study recognizes the dataset's modest size and missing attribute values. Future work includes adding strategies to improve model performance and using larger datasets for more precise predictions.

In 2020, Xue, Min, and Ma [6] discussed the increasing occurrence of type 1 diabetes in young people and emphasized the importance of early prediction to reduce the risk of delayed treatment and long-term complications. The study incorporated data from 520 individuals with diabetes or at risk of developing diabetes, ranging in age from 16 to 90. The analysis employed supervised machine-learning methods, namely SVM, Naive Bayes classifier, and LightGBM. In classification accuracy comparisons, Support Vector Machines (SVM) outperforms other methods in predicting diabetes. In 2020, Ljubic et al. [7] sought to estimate ten type 2 diabetes problems' prognosis. By anticipating these issues, focused and timely interventions can avoid or decrease their spread. The study used 2003–2011 Healthcare Cost and Utilization Project State Inpatient Databases of California data. Instead of Random Forest and Multilayer Perceptron, the study used LSTM and GRU. By evaluating the few hospitalizations between type 2 diabetes mellitus (DM2) diagnosis and related issues, the study examined the ability to predict complications. Diagnostic trials verified RNN GRU model functionality. The model correctly predicted 73% myocardial infarction and 83% chronic ischemic heart disease. Traditional models were 66%–76% accurate. The study found that hospitalizations significantly affect prediction accuracy. Hospitalizations are usually better with four than two. Deep learning models need lots of training data, and 1000 patient data yielded the best results. Prediction precision declined over time and varied by task. The study found that the RNN GRU model processed electronic medical record data better.

In 2019, Sonar et al. [8] use Machine Learning to predict diabetes, solving the challenges of frequent diagnostic appointments. This study compares Decision Trees, Artificial Neural Networks (ANN), Naive Bayes, and Support Vector Machines for categorization. SVMs are flexible with unstructured data including text, graphics, and hierarchical systems. Decision Trees are simple, yet even small data structure modifications can make them unstable. The Naive Bayes approach is resilient, although data formatting can add bias as training data grows. ANN models have high predicted accuracy, but they require a lot of computing power, especially for complicated models and large datasets. In order to improve forecast accuracy, they emphasize using modern technologies like SVM and deep learning models. Machine learning has transformed diabetes management in numerous hospital settings and addressed specific challenges, according to the research.

3 Proposed Methodology

The dataset is the foundation for diabetes prediction machine learning applications. After that, the dataset is split into training and testing sets to rigorously evaluate the chosen machine learning models. Maintaining dataset integrity requires data cleaning to correct missing or erroneous values. Then, scaling is used to standardize numerical properties to avoid scale inconsistencies that could affect machine learning approaches. Next, outliers are eliminated or modified to reduce their impact on model training. Next, XGBoost and LightGBM, which excel at diabetes prediction, are used. The final step is to evaluate accuracy measurements to choose the best algorithm. The iterative selection and evaluation of models using cutting-edge machine learning in a structured workflow ensures a powerful diabetes prediction model (Fig. 1).

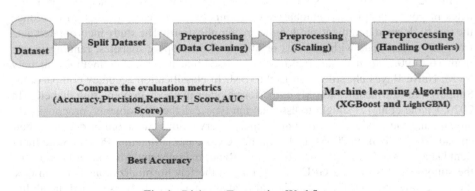

Fig. 1. Diabetes Forecasting Workflow

3.1 Data Collection

Individual physiological and demographic factors are used to predict diabetes occurrence in the diabetes dataset. The features include pregnancies, plasma glucose, diastolic blood pressure, triceps skin fold thickness, 2-h serum insulin levels, BMI, age, and a diabetes pedigree function that predicts diabetes based on family history. Dichotomous dependent variable represents diabetes (1) or absence (0). Machine learning researchers use this dataset to build predictive models for diabetes-related traits. Scientists use classification algorithms to evaluate this dataset to improve diabetes detection and treatment.

3.2 Pre-processing

Machine learning model data must be preprocessed to ensure integrity, dependability, and algorithm applicability. Data cleaning addresses dataset defects, inconsistencies, and missing values during preparation.

3.3 Data Cleaning

Data cleaning is a crucial step in the transformation of raw data into valuable insights. Data cleansing is the process of methodically finding and correcting problems in a dataset, such as missing values, duplicates, inaccuracies, and outliers. The objective is to improve the quality of the data, guaranteeing its reliability and suitability for analysis or machine learning applications. Dealing with missing values, resolving duplication, rectifying mistakes, and managing outliers are essential procedures. In addition, the process of standardizing formats, addressing incorrect category values, and eliminating extraneous information all contribute to the creation of a clean dataset. The advantages encompass greater model performance, increased interpretability, and the mitigation of biases. Although data cleansing might be challenging, it is crucial for fully harnessing the power of data in making educated decisions and developing models.

3.3.1 Scaling

Scaling is a crucial data analysis preprocessing step that standardizes measurement scales. It is crucial when working with variables having naturally fluctuating ranges or units. Scaling aims to reduce the dominance of individual features in analysis or their disproportionate influence on machine learning models due to their bigger scale.Data analysis uses Min-Max, Standard (Z-score normalization), and Robust scaling. Min-Max Scaling transforms data to [0, 1]. Standard scaling normalizes data to a mean of 0 and a standard deviation of 1. However, Robust Scaling normalizes data using the median and IQR, reducing outliers.

Scaling improves model performance, making it important. Distance computations are used in many machine learning techniques, and characteristics with different scales can skew results. Scaling ensures that every feature contributes proportionally, speeding convergence and improving model predictions.

Additionally, scaling improves comparisons and simplifies model interpretation. It ensures model coefficients or feature relevance may be compared, providing cleaner insights.The context, data properties, and analytical or modeling purpose should determine the scaling strategy. The data pretreatment pipeline needs scaling to produce more reliable and impartial statistics and machine learning results.

3.3.2 Handling Outliers

Outliers must be handled properly during data preprocessing to minimize their impact on statistical research and machine learning models. The Interquartile Range (IQR) reliably detects and handles dataset outliers.Interquartile range (IQR) is the difference between the dataset's first and third quartiles. Data points below Q1 minus the multiplier and IQR or above Q3 plus the multiplier and IQR are outliers. The multiplier is set.This method calculates the interquartile range (IQR), finds outliers using a multiplier, and manages them. Based on data qualities and research goals, handling involve changing it into better values (capping) [9].

The interquartile range (IQR) is strong at preventing extreme readings, may alter outlier thresholds with a multiplier, and maintains data integrity. By properly handling data points that differ significantly from the norm, the dataset's dependability improves, improving statistical analysis and model construction.The multiplier selection and dataset size must be considered while utilizing the IQR technique [10]. This procedure is reliable for improving data accuracy and understanding in future analysis (Fig. 2).

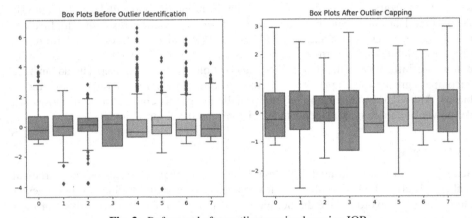

Fig. 2. Before and after outlier capping by using IQR

3.4 Models for Diabetes Prediction

3.4.1 XGBoost

XGBoost, or eXtreme Gradient Boosting, is a popular machine learning approach for its performance and scalability. In the ensemble learning framework, XGBoost is popular for classification and regression tasks due to its ability to interpret structured/tabular data [11]. XGBoost sequentially builds weak learners, usually decision trees, to develop a strong and accurate prediction model. The iterative boosting technique allows each tree to rectify earlier mistakes, creating a powerful ensemble model. Regularization makes XGBoost unique. L1 (Lasso) and L2 (Ridge) regularization variables in the objective function reduce overfitting and increase model generalization in XGBoost [12]. This is crucial for complex datasets.

Gradient boosting with decision trees (GBDT) is where XGBoost optimizes and builds trees utilizing gradients and second-order gradients. Parallel and distributed computing makes it scalable and capable of handling enormous datasets [13]. The unique ability of XGBoost to manage missing data reduces the need for extensive data imputation. The technique helps identify variables and improve interpretability by revealing feature value [14]. The approach uses column block grouping for continuous values and sparsity-aware processing for sparse data to improve computational performance. In banking and healthcare, XGBoost, a machine learning algorithm, is fast and accurate. This gives it an edge in machine learning competitions like Kaggle.

3.4.2 LightGBM

LightGBM, a cutting-edge gradient boosting framework, dominates machine learning because to its speed, scalability, and efficiency, especially with huge datasets. Light-GBM, from Microsoft, uses gradient boosting [15]. To attain accuracy and resilience, it iteratively combines weak learners, usually decision trees, to build a strong prediction model. Using leaf-wise tree development, LightGBM stands out. Instead of growing trees step-by-step, LightGBM grows trees by producing information-maximizing leaves [16]. This can improve tree structure balance and efficiency. The technique uses gradient-based optimization and histograms to find appropriate split points, improving compute efficiency and training speed [17]. LightGBM supports categorical characteristics without one-hot encoding, saving memory and speeding computations. Histogram-based learning divides continuous data into discrete bins. This method accelerates training [18]. LightGBM can process large datasets efficiently using parallel and GPU computing. LightGBM is a powerful machine learning algorithm that solves speed, scalability, and projected accuracy difficulties [19]. This choice is vital owing to its adaptability and outstanding features, which can be used for everything from data analysis to complex machine learning.

4 Result and Discussion

4.1 Performance Assessments of XGBoost with IQR

Using Interquartile Range (IQR) preprocessing to predict diabetes, the XGBoost model performed well across various criteria. With an accuracy of 0.86, the model predicted diabetes well. The precision, which evaluates positive prediction accuracy, was 0.80, suggesting that 80% of diabetes cases were correct. The model's recall was 0.81, indicating that it successfully identified 81% of genuine diabetes cases. The harmonic mean of precision and recall, the F1 Score, was 0.81, showing that the model balanced precision and recall. The model's robust class distinction was confirmed by the Area Under the ROC Curve (AUC) score of 0.80. These results show that the XGBoost model with IQR preprocessing accurately predicts diabetes. It balances precision and recollection and is selective. However, the context and application requirements must be considered when interpreting these metrics to ensure that the model matches the expected findings and healthcare environment limits (Table 1, Figs. 3, 4 and 5).

Table 1. IQR outlier detection with XGBOOST performance metrics

XGBoost with IQR Results	
Metrics	Values
Accuracy	0.86
Precision	0.80
Recall	0.81
f1_score	0.81
AUC Score	0.80

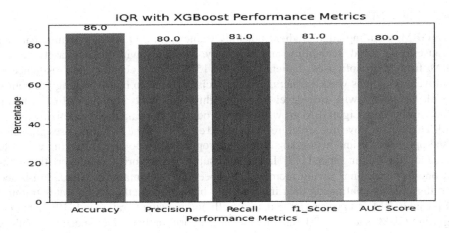

Fig. 3. IQR Outlier Detection XGBoost Performance Metrics

4.2 Performance Assessments of LightGBM with IQR

The LightGBM model and IQR preprocessing for diabetes prediction showed different performance indicators, each revealing its efficacy. With an accuracy of 0.72, the model can predict diabetes status in 72% of cases, suggesting intermediate accuracy. Precision, which assesses positive forecast accuracy, was 0.59. This shows that 59% of diabetes cases were correct. The model identified 70% of actual diabetes patients with a recall of 0.70. Precision and recall were balanced in the F1 Score of 0.64. Additionally, the Area Under the ROC Curve (AUC) was 0.78, showing strong discriminating. The diabetes prediction task may require further investigation or model improvement to satisfy healthcare goals (Table 2 and Figs. 6, 7 and 8).

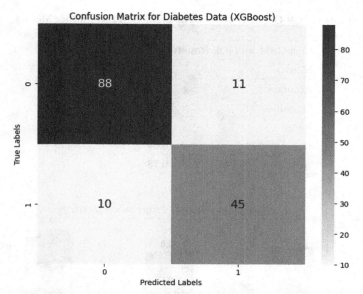

Fig. 4. Confusion Matrix for IQR Outlier Detection with XGBoost

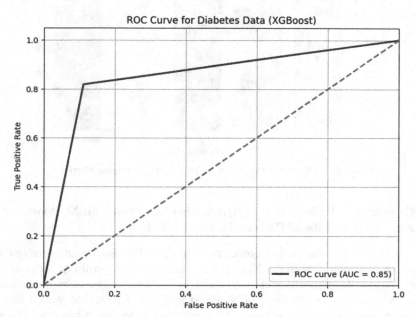

Fig. 5. ROC for IQR Outlier Detection with XGBoost

Table 2. IQR Outlier Detection with LightGBM Performance Metrics

LightGBM with IQR Results	
Metrics	Values
Accuracy	0.72
Precision	0.59
Recall	0.70
f1_score	0.64
AUC Score	0.78

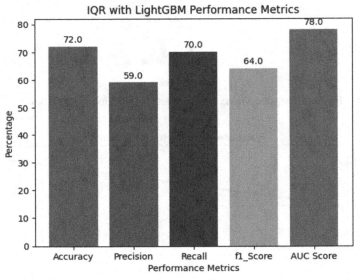

Fig. 6. IQR Outlier Detection LightGBM Performance Metrics

4.3 Comparative Performance of IQR Outlier Detection with XGBoost and LightGBM-Based Diabetes Disease Prediction

The comparative investigation of outlier detection using the interquartile range (IQR) method, in conjunction with the XGBoost and LightGBM algorithms, for predicting diabetic illness, demonstrates clear differences in performance between the two algorithms. When it comes to accuracy, XGBoost surpasses LightGBM with a superior accuracy rate of 0.86, as opposed to 0.72. Comparatively, XGBoost showcases a higher precision of 0.80, whereas LightGBM has a lesser precision of 0.59. LightGBM outperforms XGBoost in terms of recall, with a value of 0.70 compared to XGBoost 0.81. This suggests that LightGBM is more effective at identifying a greater number of true positive cases. XGBoost outperforms LightGBM in terms of F1 score, with a score of 0.81 compared to LightGBM 0.64. When it comes to discriminatory power, XGBoost

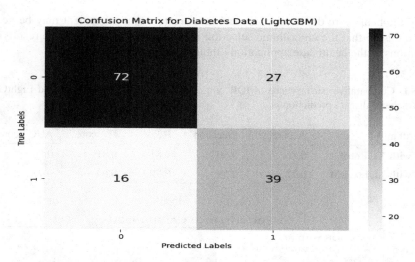

Fig. 7. Confusion Matrix for IQR Outlier Detection with LightGBM

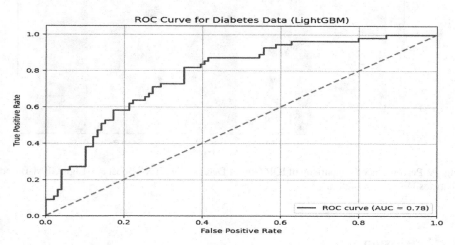

Fig. 8. ROC for IQR Outlier Detection with LightGBM

has a competitive AUC score of 0.80, although LightGBM achieves a slightly higher score of 0.78.

In summary, this analysis compares the advantages and disadvantages of XGBoost and LightGBM in predicting diabetic illness. Although XGBoost showcases exceptional accuracy, precision, and F1 score, LightGBM outperforms in recall, indicating its ability to recognize a greater proportion of true positive cases. The selection of these algorithms may be contingent upon certain criteria, such as the prioritization of precision or recall, within the healthcare domain. Additional investigation and refinement may be necessary to ensure that the algorithmic selection is in line with the anticipated results and limitations of the healthcare application (Table 3 and Fig. 9)

Table 3. Comparative performance of IQR outlier detection with XGBOOST and LightGBM based diabetes disease prediction

Algorithm	Accuracy	Precision	Recall	f1_score	AUC Score
IQR with XGBoost	0.86	0.80	0.81	0.81	0.80
IQR with LightGBM	0.72	0.59	0.70	0.64	0.78

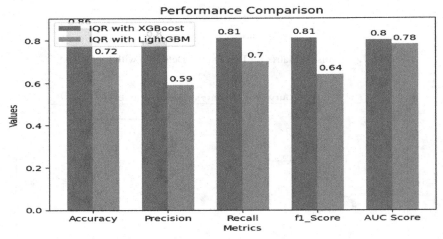

Fig. 9. Performance Comparison of IQR Outlier Detection with XGBoost and LightGBM-Based Diabetes Disease Prediction

5 Conclusion

In a nutshell, Diabetes prediction using machine learning was the main goal of this study. A thorough approach was taken to dataset preparation, preprocessing, and model validation. The dataset was used to train and evaluate machine learning models, focusing on XGBoost and LightGBM. This dataset's quality and integrity were improved by thorough data cleaning, scaling, and outlier handling, including the Interquartile Range (IQR) technique. IQR preprocessing paired with XGBoost produced durable performance metrics. With an accuracy of 0.86, it predicted diabetes well. The precision score of 0.80 indicates great accuracy in forecasting good outcomes. The model's 0.81 recall score shows its accuracy in diagnosing diabetes. A balanced F1 Score of 0.81 and an AUC score of 0.80 confirm the model's ability to distinguish positive and negative circumstances.

However, the LightGBM model with IQR preprocessing performed differently. Precision was 0.59, recall 0.70, and accuracy was 0.72, suggesting average correctness. The F1 Score, which balances precision and recall, was 0.64, showing moderate discrimination. AUC value of 0.78 indicates the model's discriminatory strength in distinguishing positive and negative events.This study addressed the essential issue of diabetes prediction utilizing advanced machine learning. Methodical preprocessing and XGBoost and LightGBM model evaluation yielded excellent findings. XGBoost accuracy, precision, and recall made it a promising diabetes predictor. However, environment and application needs should dictate model selection. This study highlights the importance of rigorous preprocessing and model validation in developing reliable and efficient diabetes prediction models.

References

1. Abhari, S., Kalhori, S.R., Ebrahimi, M., Hasannejadasl, H., Garavand, A.: Artificial intelligence applications in type 2 diabetes mellitus care: focus on machine learning methods. Healthc. Inf. Res. **25**(4), 248–261 (2019)
2. Shifrin, M., Siegelmann, H.: Near-optimal insulin treatment for diabetes patients: a machine learning approach. Artif. Intell. Med. **107**, 101917 (2020)
3. Sowah, R.A., Bampoe-Addo, A.A., Armoo, S.K., Saalia, F.K., Gatsi, F., Sarkodie-Mensah, B.: Design and development of diabetes management system using machine learning. Int. J. Telemedicine Appl. **2020**(1), 8870141 (2020)
4. Mujumdar, A., Vaidehi, V.: Diabetes prediction using machine learning algorithms. Procedia Comput. Sci. **165**, 292–299 (2019)
5. Sarwar, M.A., Kamal, N., Hamid, W., Shah, M.A.: Prediction of diabetes using machine learning algorithms in healthcare. In: 2018 24th International Conference on Automation and Computing (ICAC), pp. 1–6. IEEE (2018)
6. Xue, J., Min, F., Ma, F.: Research on diabetes prediction method based on machine learning. J. Phys. Conf. Ser. **1684**, 012062 (November 2020)
7. Ljubic, B., et al.: Predicting complications of diabetes mellitus using advanced machine learning algorithms. J. Am. Med. Inf. Assoc. **27**(9), 1343–1351 (2020)
8. Sonar, P., JayaMalini, K.: diabetes prediction using different machine learning approaches. In: 2019 3rd International Conference on Computing Methodologies and Communication (ICCMC) (2019)
9. Qiu, W., et al.: Machine learning for detecting early infarction in acute stroke with non–contrast-enhanced CT. Radiology **294**(3), 638–644 (2020)
10. Gates, A., Johnson, C., Hartling, L.: Technology-assisted title and abstract screening for systematic reviews: a retrospective evaluation of the Abstrackr machine learning tool. Syst. Rev. **7**, 45 (2018). https://doi.org/10.1186/s13643-018-0707-8
11. Wang, M.X., Huang, D., Wang, G., Li. D.Q.: SS-XGBoost: a machine learning framework for predicting newmark sliding displacements of slopes. J. Geotech. Geoenvironmental Eng. **146**(9), 04020074 (2020)
12. Torlay, L., Perrone-Bertolotti, M., Thomas, E., Baciu, M.: Machine learning–XGBoost analysis of language networks to classify patients with epilepsy. Brain Inf. **4**(3), 159–69 (2017)
13. Zhu, X., Chu, J., Wang, K., Wu, S., Yan, W., Chiam, K.: Prediction of rockhead using a hybrid N-XGBoost machine learning framework. J. Rock Mech. Geotech. Eng. **13**(6), 1231–1245 (2021)

14. Inoue, T.: XGBoost, a machine learning method, predicts neurological recovery in patients with cervical spinal cord injury. Neurotrauma Rep. **1**(1), 8–16 (2020)
15. Shehadeh, A., Alshboul, O., Al Mamlook, R.E., Hamedat, O.: Machine learning models for predicting the residual value of heavy construction equipment: an evaluation of modified decision tree, LightGBM, and XGBoost regression. Autom. Constr. **129**, 103827 (2021)
16. Gan, M., Pan. S., Chen, Y., Cheng, C., Pan, H., Zhu, X.: Application of the machine learning lightgbm model to the prediction of the water levels of the lower columbia river. J. Mar. Sci. Eng. **9**(5), 496 (2021)
17. Nemeth, M., Borkin, D., Michalconok, G.: The comparison of machine-learning methods XGBoost and LightGBM to predict energy development. In: Silhavy, R., Silhavy, P., Prokopova, Z. (eds.) Computational Statistics and Mathematical Modeling Methods in Intelligent Systems. CoMeSySo 2019 2019. Advances in Intelligent Systems and Computing, vol. 1047, pp. 208–215. Springer, Cham (2019). https://doi.org/10.1007/978-3-030-31362-3_21
18. Wang, D., Zhang, Y., Zhao, Y.: LightGBM. In: Proceedings of the 2017 In-ternational Conference on Computational Biology and Bioinformatics (2017)
19. Xia, H., Wei, X., Gao, Y., Lv. H.: Traffic prediction based on ensemble machine learning strategies with bagging and lightgbm. In: 2019 IEEE International Conference on Communications Workshops (ICC Workshops), pp. 1–6. IEEE (2019)

Enhancing Cardiovascular Health Prediction: A Machine Learning Perspective

Ratnam Dodda[✉], Abhishek Reddy Bonam, Srinidhi Sakinala, and Prithika Reddy Dendhi

CVR College of Engineering, Vastunagar, Hyderabad 501510, Telangana, India
ratnam.dodda@gmail.com

Abstract. Heart disease is a serious health concern that affects a large population worldwide. This work addresses these cardiac problems in a novel way by utilizing a variety of machine learning (ML) algorithms. Our research builds predictive models targeted at early detection and the creation of customized treatment plans by analyzing extensive patient data. It holds the key to a major breakthrough in cardiovascular care, allowing for early detection and personalized prevention strategies. The main goal is to provide innovative resources to medical professionals so that they can anticipate and treat heart diseases proactively and achieve better results.

Keywords: Heart Diseases · Decision Tree · KNN · Logistic Regression · SVM

1 Introduction

Machine Learning(ML), a subset of Artificial Intelligence(AI), revolves around crafting statistical models that empower automated systems to execute functions independently, devoid of direct human involvement or intricate programming [1,2]. It aims to equip computers with the power to learn from data, enabling them to predict outcomes and make informed decisions. Machine learning algorithms use statistical inference and patterns instead of explicit programming to improve with time as they are exposed to more data [3,4]. Heart ailments persist as a significant health concern that affects millions of people each year. Urbanization and modernization processes are often linked to an increase in non-communicable diseases, especially cardiovascular conditions [5,6]. Machine learning (ML) algorithms have become effective instruments in the healthcare field to handle the complexity of heart diseases [5,7]. These algorithms use broad datasets-such as genetic data, medical imaging, and patient records-to examine correlations and patterns that may go undetected by conventional techniques.

R. Dodda, A. R. Bonam, S. Sakinala and P. R. Dendhi—Contributing authors.

There are many different uses for machine learning in cardiovascular health. Algorithms that can identify subtle patterns in patient data are useful for risk assessment and diagnosis [8,9]. With the help of genetic predispositions, lifestyle factors, and medical history, ML models can predict a patient's risk of developing heart disease. This makes it possible to provide individualized risk assessments and prompt interventions. With its skill at deciphering complex patterns and analyzing large datasets, machine learning is establishing itself as a vital tool in the continuous fight against cardiovascular illnesses [10,11].

2 Methodology

Our primary goal with the suggested approach is to facilitate the early identification of cardiovascular diseases by predicting its occurrence within a short timeframe. Our approach involves leveraging various data mining techniques and algorithms, namely k Nearest Neighbor (KNN), Decision tree, Logistic regression, Support vector machine (SVM). These algorithms utilize health parameters to make predictions. The analysis of the data is conducted within Google Colab, a cloud-based platform. It provides a collaborative environment for writing code, performing data analysis, and implementing machine learning algorithms. Additionally, the platform facilitates the creation of live code, enables visualization, supports data processing, and aids in plotting graphs [11,12].

2.1 Data Collection

We utilized a dataset sourced from the UCI Machine Learning Repository for our analysis focused on predicting heart disease.
https://archive.ics.uci.edu/

2.2 Features

This database contains 14 attributes listed below:
age
sex
chest pain type (4 values)
resting blood pressure
serum cholesterol in mg/dl
fasting blood sugar > 120 mg/dl
resting electrocardiographic results (3 values)
maximum heart rate achieved.
exercise-induced angina.
old peak (ST depression induced by exercise relative to rest)
the slope of the peak exercise ST segment
number of major vessels (colored by fluoroscopy) thal (3 values)

2.3 Data Splitting

The dataset is split into two subsets: a training subset that includes 80% of the data for model training, and a testing subset with the remaining 20% for assessing the model's performance. To improve the data's uniformity, we applied normalization and scaling methods, which allowed for a linear transformation.

2.4 Algorithms

2.4.1 KNN

The KNN algorithm is universally recognized in healthcare and predictive analytics, especially in heart disease prediction. It is a frequently used tool for identifying patterns and generating predictions based on similarities between data points because of its adaptability and resilience. The algorithm is trained in the context of heart disease prediction using a dataset enhanced with relevant health indicators, age, blood pressure, cholesterol levels, and other features. Using these characteristics, the algorithm finds complex correlations and patterns that are essential for forecasting the development of heart disease [13]. As a non-parametric algorithm, KNN classifies a given data point based on the predominant class among its 'k' closest neighbors. To forecast the likelihood of a person developing heart disease, KNN examines past patient data and applies knowledge from comparable cases [14,15].

The spatial separation between the data points x_i and x_{test} in a KNN algorithm can be computed using a distance measure such as the Euclidean distance.:

$$\text{distance}(x_i, x_{\text{test}}) = \sqrt{\sum_{j=1}^{n}(x_{i,j} - x_{\text{test},j})^2} \qquad (1)$$

2.4.2 Decision Tree

The Decision Tree method is a widely utilized machine learning strategy that is often employed in predicting the occurrence of coronary artery disease. By frequently segmenting the data into groups according to the most important characteristics, this technique creates a tree-like structure that aids in decision-making. In the realm of heart disease, decision trees are employed to estimate the probability of heart-related complications by analysing patient-specific information [16]. Decision trees are particularly useful in healthcare settings because they are transparent and easy to understand. They work well with a variety of patient datasets because they can handle numerical and categorical data with ease. Decision trees provide distinct decision pathways for predicting the risk of heart disease by capturing complex relationships within the data through the recursive evaluation of features [17]. The most informative features are chosen to partition the feature space at each node, optimizing for the best split to maximize information gain or minimize impurity [18,19].

Data: Training set $D = \{(x_i, y_i)\}$, where x_i is the feature vector and y_i is the label, Test instance x_{test}, Number of neighbors k
Result: Predicted label $\hat{y}$ for x_{test}
for $i = 1$ *to* $|D|$ **do**
  | Calculate the distance between x_{test} and x_i: $d_i = \text{distance}(x_{\text{test}}, x_i)$;
end
Sort the distances d_i in ascending order and get the indices of the first k neighbors;
k-nearest neighbors: $N = \{i_1, i_2, \ldots, i_k\}$;
for $j = 1$ *to* k **do**
  | Accumulate the labels of the k-nearest neighbors: $C[y_{i_j}]\mathrel{+}= 1$;
end
Find the label with the maximum count: $\hat{y} = \arg\max_c C[c]$;
Output: Predicted label $\hat{y}$
Algorithm 1: K-Nearest Neighbor Algorithm

Data: Training set $D = \{(x_i, y_i)\}$, where x_i is the feature vector and y_i is the label
Result: Decision Tree model
if *stopping criterion is met* **then**
  | **return** *Leaf node with label determined by majority of y_i*;
end
Select the best feature f and split point θ using a criterion (e.g., Gini index, information gain);
Create a decision node with test (f, θ);
Split the data into subsets D_{left} and D_{right} based on the decision node;
Recursively apply the algorithm to D_{left} and D_{right} to create subtrees;
Output: Decision Tree model
Algorithm 2: Decision Tree Algorithm

The formula for entropy (H) is given by:

$$H(D) = -\sum_{i=1}^{c} p_i \log_2(p_i) \qquad (2)$$

The information gain (IG) for a feature A is given by:

$$IG(D, A) = H(D) - \sum_{v \in \text{values}(A)} \frac{|D_v|}{|D|} \cdot H(D_v) \qquad (3)$$

where:

- D_v is the subset of D for which feature A has value v,
- values(A) is the set of possible values for feature A.

2.4.3 Logistic Regression

In the realm of predictive analytics, the logistic regression algorithm is a potent and frequently used technique, especially when it comes to the medical field's use of it for cardiac disease prediction. It is used to simulate the likelihood that a person with heart disease will have certain input features. The logistic function, transforms a linear combination of features and their corresponding weights into a probability value that ranges from 0 to 1. The model, which is well-known for its comprehensibility and ability to offer probability estimates, is ideal for circumstances in which knowing the probability of an event is crucial, like heart disease [20, 21].

Data: Training set $D = \{(x_i, y_i)\}$, where x_i is the feature vector and y_i is the label
Result: Learned logistic regression parameters
Initialize weights $\mathbf{w}$ and bias b;
Set learning rate α;
while *not converged* **do**
  Calculate the logits: $z_i = \mathbf{w}^T x_i + b$ for all i;
  Calculate the predicted probabilities: $p_i = \frac{1}{1+e^{-z_i}}$ for all i;
  Calculate the log-likelihood: $L(\mathbf{w}, b) = \sum_{i=1}^{|D|} y_i \log(p_i) + (1 - y_i) \log(1 - p_i)$;
  Calculate the gradients: $\frac{\partial L}{\partial \mathbf{w}}, \frac{\partial L}{\partial b}$;
  Update weights: $\mathbf{w} \leftarrow \mathbf{w} - \alpha \frac{\partial L}{\partial \mathbf{w}}$;
  Update bias: $b \leftarrow b - \alpha \frac{\partial L}{\partial b}$;
end
Output: Learned logistic regression parameters $\mathbf{w}$ and b
Algorithm 3: Logistic Regression Algorithm

The logistic regression model utilizes the logistic function to estimate the likelihood of a binary outcome. It is given by:

$$\sigma(z) = \frac{1}{1+e^{-z}} \qquad (4)$$

The logistic regression model may thus be articulated as:

$$P(Y = 1|\mathbf{x}) = \sigma(\mathbf{w}^T \mathbf{x} + b) \qquad (5)$$

2.4.4 SVM

One powerful and adaptable machine learning algorithm used for heart disease prediction is Support Vector Machines (SVM). SVM can be particularly useful for overseeing complex non-linear relationships in data, and it may be employed to categorise patients according to parameters such as systolic pressure, age and lipid profiles. Its objective is to construct the most optimal hyperplane that segregates instances belonging to diverse classes. The SVM algorithm aims to identify a hyperplane that maximizes the distance, or margin, between the hyperplane itself and the nearest data point from each class. This approach ensures optimal separation between different classes of data points. The algorithm's ability to effectively generalise to previously unseen data depends on this margin. When data cannot be separated linearly, SVM applies the kernel trick, which converts the feature space into a space of greater dimensionality where a hyperplane can effectively divide the classes [22,23].

Data: Training set $D = \{(x_i, y_i)\}$, where x_i is the feature vector and y_i is the label
Result: Learned SVM parameters
Initialize weights **w** and bias b;
Set regularization parameter C;
Set learning rate η;
while *not converged* **do**
  Calculate the decision function: $f(\mathbf{x}_i) = \mathbf{w}^T \mathbf{x}_i + b$ for all i;
  Calculate the hinge loss: $L(\mathbf{w}, b) = \frac{1}{2}\|\mathbf{w}\|^2 + C \sum_{i=1}^{|D|} \max(0, 1 - y_i f(\mathbf{x}_i))$;
  Calculate the gradients: $\nabla_{\mathbf{w}} L, \frac{\partial L}{\partial b}$;
  Update weights: $\mathbf{w} \leftarrow \mathbf{w} - \eta(\nabla_{\mathbf{w}} L + C \sum_{i=1}^{|D|} -y_i \mathbf{x}_i)$;
  Update bias: $b \leftarrow b - \eta \frac{\partial L}{\partial b}$;
end
Output: Learned SVM parameters **w** and b
  Algorithm 4: Support Vector Machines (SVM) Algorithm

The decision function for a linear SVM is given by:

$$f(\mathbf{x}) = \mathbf{w}^T \mathbf{x} + b \qquad (6)$$

The SVM objective function with a hinge loss and L2 regularization is given by:

$$L(\mathbf{w}, b) = \frac{1}{2}\|\mathbf{w}\|^2 + C \sum_{i=1}^{|D|} \max(0, 1 - y_i f(\mathbf{x}_i)) \qquad (7)$$

3 Results and Discussion

In our exploration of heart disease prediction through machine learning, the utilization of Python proves instrumental for several reasons. The extensive use of Python in data science can be credited to its comprehensive libraries and the active support from its community. Leveraging libraries like scikit-learn, TensorFlow, and PyTorch within the Python ecosystem, we benefit from a diverse range of tools for implementing, testing, and refining machine learning algorithms.

Accuracy is an evaluation metric that quantifies the overall precision of a classification model. It is given by:

$$\text{Accuracy} = \frac{TP + TN}{TP + TN + FP + FN} \tag{8}$$

The F1 score is the harmonic average of precision and recall, offering a balanced perspective between these two parameters. It is calculated as:

$$\text{F1 Score} = \frac{2 \times \text{Precision} \times \text{Recall}}{\text{Precision} + \text{Recall}} \tag{9}$$

Precision is defined as the proportion of true positive predictions relative to the sum of all positive predictions, while recall is the proportion of true positive predictions relative to the total number of actual positive instances.

$$\text{Precision} = \frac{TP}{TP + FP} \tag{10}$$

$$\text{Recall (Sensitivity)} = \frac{TP}{TP + FN} \tag{11}$$

The depicted figure illustrates the results obtained through the implementation of the KNN algorithm. Our Python implementation of this approach yielded a classification accuracy of 70% and an F1 score of 71%, indicating the effectiveness in accurately categorizing the data (Fig. 1).

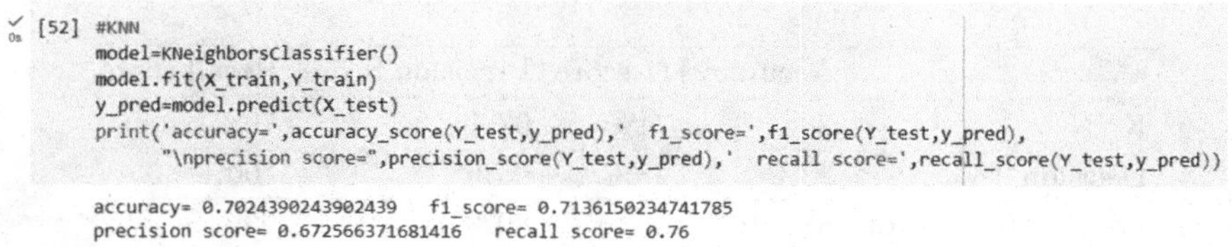

```
[52] #KNN
model=KNeighborsClassifier()
model.fit(X_train,Y_train)
y_pred=model.predict(X_test)
print('accuracy=',accuracy_score(Y_test,y_pred),' f1_score=',f1_score(Y_test,y_pred),
      "\nprecision score=",precision_score(Y_test,y_pred),' recall score=',recall_score(Y_test,y_pred))

accuracy= 0.7024390243902439    f1_score= 0.7136150234741785
precision score= 0.672566371681416    recall score= 0.76
```

Fig. 1. KNN Results

The depicted figure illustrates the results obtained through the utilization of the decision tree algorithm. Our Python implementation of this approach yielded a 98% accuracy rate and an f1 score of 98% in effectively categorizing. Building upon these outcomes, the decision tree algorithm demonstrated remarkable precision in its classification performance (Fig. 2).

The depicted figure illustrates the results obtained through the application of the logistic regression algorithm. Our implementation of this approach in Python yielded an accuracy of 85% and an F1 score of 86%, showcasing the algorithm's proficiency in accurately categorizing instances (Fig. 3).

The depicted results illustrate the application of the SVM algorithm. Our Python implementation of this technique yielded an accuracy of 73% and an

```
[49] #Decision Tree
    model=DecisionTreeClassifier()
    model.fit(X_train, Y_train)
    y_pred=model.predict(X_test)
    print('accuracy=',accuracy_score(Y_test,y_pred),' f1_score=',f1_score(Y_test,y_pred),
          "\nprecision score=",precision_score(Y_test,y_pred),' recall score=',recall_score(Y_test,y_pred))

    accuracy= 0.9804878048780488    f1_score= 0.9803921568627451
    precision score= 0.9615384615384616   recall score= 1.0
```

Fig. 2. Decision Tree Results

```
[44] #Logistic Regression
    model = LogisticRegression()
    model.fit(X_train, Y_train)
    y_pred=model.predict(X_test)
    print('accuracy=',accuracy_score(Y_test,y_pred),' f1_score=',f1_score(Y_test,y_pred),
          "\nprecision score=",precision_score(Y_test,y_pred),' recall score=',recall_score(Y_test,y_pred))

    accuracy= 0.8536585365853658    f1_score= 0.8611111111111112
    precision score= 0.8017241379310345   recall score= 0.93
```

Fig. 3. Logistic Regression Results

```
[47] #SVM
    model=svm.SVC()
    model.fit(X_train,Y_train)
    y_pred=model.predict(X_test)
    print('accuracy=',accuracy_score(Y_test,y_pred),' f1_score=',f1_score(Y_test,y_pred),
          "\nprecision score=",precision_score(Y_test,y_pred),' recall score=',recall_score(Y_test,y_pred))

    accuracy= 0.7317073170731707    f1_score= 0.7639484978540773
    precision score= 0.6691729323308271   recall score= 0.89
```

Fig. 4. SVM Results

Table 1. Accuracy Table

	Accuracy	f1_score	Precision Score	Recall score
KNN	70%	71%	67%	79%
Decision Tree	98%	98%	96%	100%
Logistic Regression	85%	86%	80%	93%
SVM	73%	76%	67%	89%

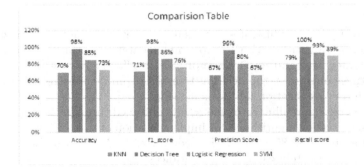

Fig. 5. Comparision Table

f1 score of 76%, indicating the effectiveness in accurately categorizing. Building upon these findings, it is evident that the SVM algorithm demonstrates proficiency in classification tasks (Fig. 4 and Table 1).

The table and chart demonstrate four models' performance metrics. The metrics include accuracy, precision, f1-score, and recall score. Higher values in the scores, which are displayed as percentages, denote superior performance (Fig. 5).

4 Conclusion

In conclusion, predictive models for cardiovascular disease which employ classification algorithms, highlight the critical importance of addressing cardiovascular conditions due to their potentially life-threatening consequences. Notable outcomes were obtained from an organised assessment of these models with an emphasis on their accuracy rates. Interestingly, the Decision Tree model outperformed KNN (70%) and SVM (73%), with an accuracy rate of 98%, while Logistic Regression only managed an accuracy rate of 85%. While the selection of a suitable model is contingent upon numerous factors such as interpretability, scalability, computational resources, and data characteristics, decision tree excelled on the dataset, demonstrating remarkable predictive power. This emphasises the potential of using massive health-related data to train these models.

References

1. Mitchell, T.M.: Machine learning (1997)
2. Sarker, I.H.: Machine learning: algorithms, real-world applications and research directions. SN Comput. Sci. **2**(3), 160 (2021)
3. Domingos, P.: A few useful things to know about machine learning. Commun. ACM **55**(10), 78–87 (2012)
4. Bolukbasi, T., Chang, K.-W., Zou, J.Y., Saligrama, V., Kalai, A.T.: Man is to computer programmer as woman is to homemaker? Debiasing word embeddings. Adv. Neural Inf. Process. Syst. **29** (2016)
5. Ruan, Y., et al.: Cardiovascular disease (CVD) and associated risk factors among older adults in six low-and middle-income countries: results from sage wave 1. BMC Public Health **18**(1), 1–13 (2018)
6. Nabel, E.G.: Cardiovascular disease. N. Engl. J. Med. **349**(1), 60–72 (2003)
7. Gaziano, T., Reddy, K.S., Paccaud, F., Horton, S., Chaturvedi, V.: Cardiovascular disease. Disease Control Priorities in Developing Countries. 2nd edition (2006)
8. Nayyar, A., Gadhavi, L., Zaman, N.: Machine learning in healthcare: review, opportunities and challenges. Machine Learning and the Internet of Medical Things in Healthcare, 23–45 (2021)
9. An, Q., Rahman, S., Zhou, J., Kang, J.J.: A comprehensive review on machine learning in healthcare industry: classification, restrictions, opportunities and challenges. Sensors **23**(9), 4178 (2023)
10. Subramani, S., et al.: Cardiovascular diseases prediction by machine learning incorporation with deep learning. Front. Med. **10**, 1150933 (2023)
11. Averbuch, T., et al.: Applications of artificial intelligence and machine learning in heart failure. Eur. Heart J. Digital Health **3**(2), 311–322 (2022)

12. Mathur, P., Srivastava, S., Xu, X., Mehta, J.L.: Artificial intelligence, machine learning, and cardiovascular disease. Clin. Med. Insights: Cardiol. **14**, 1179546820927404 (2020)
13. Kumar, P.K., Raghavendra, C., Dodda, R., Shahebaaz, A.: A novel approach to strengthening web-based cloud services: two-factor access control. In: E3S Web of Conferences, vol. 472, pp. 02001 (2024). EDP Sciences
14. Jabbar, M.: Prediction of heart disease using k-nearest neighbor and particle swarm optimization. Biomed. Res. **28**(9), 4154–4158 (2017)
15. Rahmat, D., Putra, A.A., Setiawan, A.W., et al.: Heart disease prediction using k-nearest neighbor. In: 2021 International Conference on Electrical Engineering and Informatics (ICEEI), pp. 1–6. IEEE (2021)
16. Ratnam, D., HimaBindu, P., Sai, V.M., Devi, S.R., Rao, P.R.: Computer-based clinical decision support system for prediction of heart diseases using naïve bayes algorithm. Int. J. Comput. Sci. Inf. Technol. **5**(2), 2384–2388 (2014)
17. Dodda, R., Maddhi, S., Thuraab, M.S., Reddy, A.N., Chandra, A.S.M.: NLP-driven strategies for effective email spam detection: A performance evaluation. In: 2023 International Conference on Sustainable Communication Networks and Application (ICSCNA), pp. 275–279. IEEE (2023)
18. Reddy, V.S.K., Meghana, P., Reddy, N.S., Rao, B.A.: Prediction on cardiovascular disease using decision tree and naïve bayes classifiers. In: Journal of Physics: Conference Series, vol. 2161, p. 012015 (2022). IOP Publishing
19. Agbemade, E.: Predicting heart disease using tree-based model (2023)
20. Zhang, Y., Diao, L., Ma, L.: Logistic regression models in predicting heart disease. In: Journal of Physics: Conference Series, vol. 1769, p. 012024 (2021). IOP Publishing
21. Ambrish, G., Ganesh, B., Ganesh, A., Srinivas, C., Mensinkal, K., et al.: Logistic regression technique for prediction of cardiovascular disease. Glob. Transitions Proc. **3**(1), 127–130 (2022)
22. Alty, S.R., Millasseau, S.C., Chowienczyc, P., Jakobsson, A.: Cardiovascular disease prediction using support vector machines. In: 2003 46th Midwest Symposium on Circuits and Systems, vol. 1, pp. 376–379 (2003). IEEE
23. Shah, S.M.S., Shah, F.A., Hussain, S.A., Batool, S.: Support vector machines-based heart disease diagnosis using feature subset, wrapping selection and extraction methods. Comput. Electr. Eng. **84**, 106628 (2020)

Measuring AI's Impact on HR Strategies and Operational Effectiveness in the Era of Industry 4.0

K. Devi[1], Jagendra Singh[2(✉)], Neha Garg[2], Mohit Tiwari[3], Nazeer Shaik[4], and Muniyandy Elangovan[5]

[1] Department of Commerce, DAV Autonomous College, Titilagarh, India
[2] School of Computer Science Engineering and Technology, Bennett University, Noida, India
jagendrasngh@gmail.com
[3] Department of Computer Science and Engineering, Bharati Vidyapeeth's College of Engineering, Delhi, India
[4] Department of Computer Science and Engineering, Srinivasa Ramanujan Institute of Technology – Autonomous, Anantapur, India
[5] Department of Biosciences, Saveetha School of Engineering, Saveetha Institute of Medical and Technical Sciences, Chennai, India

Abstract. This study examines the transformative impact of automation-training programs on human resources strategies and operational effectiveness in the Industry 4.0 era. Through an in-depth analysis of training-performance metrics before and after including knowledge acquisition, skill proficiency, increased productivity, and product quality, the report describes the quantifiable benefits of investment in employee development initiatives designed specifically for automation technologies. Surveys of different groups of respondents show that having undergone formal training programs with workshop practice, peer mentoring, collaborative study groups, and virtual reality-based simulations, workers gain a more profound understanding of automation concepts. They will have increased skill in operating equipment, with a consequently large leap in quality and productivity for the organization. Furthermore, it was observed that cycle times and equipment downtimes fell appreciably across the board post-training, suggesting improved operational efficiency and more streamlined production processes. This research points out the immense significance of training for automation projects led by HR managers in enabling employees to meet increasing demand for automation in the workplace, and this may well contribute greatly to a continuing learning society that is favorable to innovation--the basis of success in an Industry 4.0 digital world.

Keywords: Automation training · Industry 4.0 · Human resource strategies · Operational effectiveness · Employee development

1 Introduction

Industrial 4.0 is rapidly changing and thanks to technological advancements and digital transformation, all organizations are looking to optimize their operations processes and hold their ground in the world market. Automation technologies have significant implications for the reconfiguration of traditional manufacturing processes and fault diagnosis. The extent of the success in adopting automation tools depends not just on technological investments but also on the preparation and capability of the workforce in utilizing these technologies effectively. Thus, the research suggest that HR strategies for the training and cultivation of employees are significant factors in the field of Industry 4.0 [1, 2].

This research will be an attempt to define the links between automation training and HR strategies in combination with productivity and effective operations in the age of Industry 4.0. The purpose of the investigation is to provide practical information to HR initiatives dealing with the challenges presented by today's increasingly industrialized environment. When considering the effect of automation training programs on the development of employees' skills, productivity and overall operational effectiveness [3].

Automation technology has always been a mixed blessing for organizations in many fields. Automation could, on the one hand, raise efficiency, save money and speed up innovation. But to gain the full benefits these qualities must be possessed also by workers who are expert in running automatic systems as well as making them. HR departments, therefore, have to implement all sorts of training and development programs. They need to give employees the know-how to succeed in environments that operate entirely by auto-pilot [4].

In this paper, we wish to investigate how automation training programs affect HR strategies and increase operational efficiency. In this study, by comparing changes in performance measures before and after training (including skill mastery, measurable productivity gains and improvements in product quality), the tangible impact of automation training on an enterprise's performance will be retrieved. A survey that takes in the entire spectrum of such exchanges will yield some useful working guidelines. This work can serve HR management personnel and business leaders, and policy makers to answer basic questions like "what's going on out there" in the face of and as a result the digital revolution sweeping industry today [5, 6].

2 Literature Review

A literature review shows a comprehensive overview of research and academic/scholarly works on automation training, HR strategy, and operational effectiveness in the context of Industry 4.0. As companies try to prepare for the challenges and opportunities presented by advanced automation technologies, automation training programs have become an essential part of their organization strategy [7, 8]. There are a number of training methods involved in these programs, such as hands-on workshops, virtual reality simulations, and software-based learning module to develop employee operational and managerial capabilities. Research shows the importance of molding training programs to suit the specific needs of employees, and making them align with organizational goals so that they work optimally [9, 10].

The literature even shows us that the strategic function of HR is to boost and run automation training courses. In helping to determine workforce success and improving HR programs and helping workers do so through becoming skilled labor in the most needed areas, human resource experts are of significance. HR policies designed for continuous improvement in learning and skills enhancement are what will enable manning the production line quite effectively for Industry 4.0. Yet research has proved that it is sound communication and change-management skills which are essential for making automation technologies operational. In this way, companies can also introduce them into organizational life with less difficulty [11, 12].

It has also been investigated how automation training changes the operational efficiency of organization [13]. According to research, well-structured and fully implemented training programs can greatly improve worker skill proficiency and job satisfaction, as well as job performance Studies have proven scientific that people can get rid of many problems. Organizations hoping to raise efficiency and lighten the load of resources through fewer errors and operational errors as well as information inefficiencies. Finally, education and vocational courses for automation have also resulted in improvements in product quality and/or increased competitiveness: customer satisfaction. This shows its strategic importance for achieving organizational performance [14].

In addition to the advantages of automated training, disadvantages and barriers were also acknowledged in the literature. These might include inertia, lack of organizational support or resource constraints, which not only directly hinder the success of training programs but also slow the adoption of Automation Technologies. As for automation, meeting these barriers requires implementing a comprehensive solution that includes signs of inclusive validation internally supported and committed management give substance to the words [11]. This text must be read in conjunction with the Evaluation and Feedback System to Provide Assurances of Authority's Promise.

Automation training is essential in an Industry 4.0 workplace where HR strategies and operational effectiveness are challenged. Demands for workers to possess new or developing skills will exponentially increase, given the potential automation to reshape society through digital inter-brand competition that follows some concept of Global Village. Companies that invest in human resources development and build a learning culture inside themselves can survive and even thrive in the world of tomorrow.

3 Methodology

TechPro Production is an automation technology company integrating robotics and CNC machines. For companies like TechPro Production, such integration of automation technology means both opportunities and challenges. Despite its promise of improvements in efficiency and productivity, these systems can only be successful if the workforce is willing to adopt and has the capability to use them.

Up to now, we have applied HR methods in TechPro Production that will help bear the automation. Because both small and large companies are shedding workers, the forward-looking manufacturer TechPro Production is automating. We have installed pneumatic robots for pick-and-place tasks on the production line and introduced CNC machines for high-precision machining. One feature that makes pneumatic robots different from mechanical ones is that their gripping end effector is designed specifically for holding and manipulating objects. To get CNC machines to perform drilling, cutting, and milling accurately keeping a smooth robot path, operators must be adept at reading both G-code and M-code instructions. Nonetheless, TechPro Production recognizes that without a fully mobilized workforce it will not be able to gain full benefit from automation.

To ensure that automation know-hows blend well, TechPro Production's HR department has designed special training programs suited to the needs of the company's employees who run and maintain robots and CNC machines. The training programs include both theory lessons and practical exercises as well as continuing opportunities to learn. Personnel who are tasked with operating pneumatic robots need to grasp the basic principles of robot operation, programing and troubleshooting. Safety precautions are also reviewed. Moreover, with the aid of pneumatic technology, how one might work with solenoid-operated directional control valves and electric motors to achieve technical mastery over a balanced diet. Training modules also cover rotational movement for best positioning in the robot base. Similarly, employees responsible for running a computer numerical control (CNC) machine receive an education in reading and understanding G-code and M-code instructions. They also need to recognize machine functions and perform maintenance routines. In practice, operators choose the appropriate tools, and then they compare them so that by selecting their own tooling carefully they become expert at setting up machining jobs.

3.1 Understanding Employee Knowledge and Training Requirements

We conducted semi-structured interviews in order to understand how much our employees know about automation technologies and what their training needs may be. By focusing on open-ended conversation rather than structured questioning, these interviews allowed personnel to express their histories and trends in knowledge about the area of business we were studying -- rather than trotting out trite cliches or total misunderstandings. The responses prompted them to think about the educational actions that had surrounded these impressions, as well as any changes they expected (or feared). Questions appropriately chosen for the audience's positions drew out information useful to the panel in training needs for pneumatically-operated robots and CNC machines.

In addition to interviews, the survey instruments such as Likert scales and multiple-choice questions were used to fill in gaps and get a more detailed picture. We had a standardized instrument on which participants ticked off just how good they believed themselves at operating automatic technology or related tasks. The questionnaire was geared specifically for this purpose. Moreover, document analysis also helped supplement insights obtained from interviews and surveys. This involved a systematic review of existing training materials, standard operating procedures (SOPs), and related technical documentation about pneumatic-operated robots, and CNC machines. By studying each training manual, equipment manual, and internal memo separately the research team learned much about what has been tried before. A mix of interviews, surveys, and document reviews painted a multifaceted picture of what employees need to know about automation technology and what they know. With each method having a unique advantage, the data from one method could be triangulated with the other. Interviews provided qualitative insights. Combining findings from interviews along with numbers obtained in surveys and then supplementing these with readings of official documents revealed to us what the employees at our company as a whole really knew, as well as where we might best intervene.

3.2 Training Methods Implemented in the Current Research

The automation technology implemented training program aimed at imparting the knowledge, skills, and self-confidence to employees that were necessary for firmly mastering the skills of operating CNC machines and pneumatic robots within the organization. In this unique course, not only do participants gain an understanding of how these machines are programmed, they also participate in four-hour learning formats which include live workshops filmed with 360-degree video -and-simulated scenarios viewed through VR glasses. That task is assigned to experienced students who have completed the course. To resolve problems or help them progress in their work, mentors accompany trainees. Automation Studio software will be used in the course after this. It will help learners and employees to our company. This program ultimately aims to increase efficiency by fostering cooperation and promoting knowledge sharing. It also prepares everybody for the coming social practices in automation. The overall training given in the research are shown in Fig. 1.

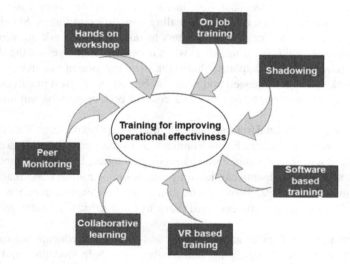

Fig. 1. Various training given in this research

To provide participants with practical experience in operating CNC machines and pneumatically driven robots, the training program offers hands-on workshops and demonstrations. Participants will have the opportunity to learn about various types of equipment while using these interactive sessions. Now what exactly are the different machine types -why they do certain tasks, and how they operate. Yours is to know everything about them. At the same time it can save lives. Participants should do this under the guidance of experienced teachers with practical training. Through hands-on exercises To make a solid foundation for other forms of training -- equipment operation, crash analysis safety norms will all be involved in that kind course.

Mentoring between peers will have interactive discussions, group activities and skill-sharing exercises for participants to promote collaboration and knowledge sharing effectively. Mentors will encourage active participation and develop a positive learning environment. They will give constructive feedback to help people develop their skill and confidence and encourage participants who are shy or self-conscious to take on a more active role. In closely working with their peers, people can gain valuable insights, improve their problem-solving skills, and foster strong relationships with their professional partners. These activities work to strengthen the foundation for recurring collaboration and contact.

As the initial training continues, personnel participating in the program will switch over to on-the-job learning and opportunities for shadowing. Gaining practical experience is just as important. During production activity led by experienced persons, participants not only apply their new knowledge but also practice equipment operation, reconciliation and trouble-shooting skills to solve problems. Likewise, Through hands-on experience and watching with your own eyes, participants grow more confident in the skills they have honed and become adept at their jobs in performing them accurately and efficiently. On-the-job training brings chances to put things to practical use. Also, you can receive immediate feedback from mentors on the spot and enjoy the real-world challenges and successes that come from dealing with different things. Standing on its own merits, mentors will prompt participants to take initiative; ask questions as the need arises and seek guidance from others when necessary. Better yet are the shadowing experiences in store for participants. Observing their practice in practice, participants can gain a look at things done well and what can be done well. Best practices and keen observation in turn are the touchstones of success for both operations and maintenance personnel.

To supplement traditional training methods, trainees will experience VR-based training exercises, presenting them with environments and situations conducive to immersive learning. There will be VR simulations like living factories, lifelike equipment operation scenes, so that participants can learn by practicing operating CNC machines and bonding fixtures with robots. Through true-to-life tutorials, manual exercises and computer game-like challenges students will learn to think critically, make decisions by themselves.

VR simulations will offer an immersive and interactive educational experience, which together with real-time advice and feedback can help students build habits of intelligent action and avoid common errors. Teachers will preside over sessions and watch the progress of their students, providing help when necessary. Participants will be able to pass through many different virtual scripts, and by practicing in a risk-free environment, they will have many opportunities to make mistakes and learn from them in security. But also, productivity increases when people are emotionally engaged because they are more motivated and remember what they learn better.

Participants will receive support and guidance from instructors who can help answer questions or explain the material in order to assure understanding at all levels in a sort of practical way. The required number of work stations must be available for the class size, and the company applies to offer this course only in a traditional classroom environment. Before being able to revise the 16-h version of the introduction to software programming language, please complete all software sessions and pass the final exam.

4 Result and Discussion

The success of HR-led initiatives was determined by several metrics One such method was through a Knowledge Assessment taken both before and after the training to determine how well people understood automation technology. Participants' knowledge acquisition and retention were scored from 0–10; the average was found to be around five. This test allowed us to verify to what extent knowledge relevant to human resources management automation technology has been disseminated among the staff. Also it provided them with some new ways of seeing and thinking about things.

In addition, Skill Proficiency was measured before and after the training to gauge the employees' actual capacity for running such devices as CNC machine tools or air-powered robots. As with the Knowledge Assessment, participants were rated on a scale of 0–10 according to their ability to perform tasks safely and accurately. Skill proficiency The HR team was able to evaluate whether the training program was successful in teaching employees how to run automation systems effectively in production line conditions.

Moreover, data post-training that productivity and product quality had brought about a computational improvement as well as control, was observed among different five groups before and after intervention. The company also saw an increase in productive work hours, falling idle time and reduced maintenance as the content improved as well, just to mention a few indicators of increased productivity. Similarly, the quality of products was improved in direct proportion to error rates and defects decreasing. This showed that programs with a clear purpose and vigorous performance brought about significant improvement in all circles.

After training courses, operational-standards (i.e. production output amounts, cycle time durations, and equipment downtime times) underwent a fresh examination in order to assess the improvements made to operational efficiency and effectiveness. The HR team was able to use these metrics as a guide for judging whether the training program had really resulted in increased throughput and had reduced cycle times and equipment problems in machinery.

Performance scores of 30 employees before and after a new automation training are outlined in Figs. 2, 3 and 4. This report focuses on knowledge assessment and skill proficiency as well as productivity improvement and product quality improvement.

Before training, employee knowledge assessment scores ranged from 5 to 8, indicating varying levels of understanding of automation technologies. Similarly, skill proficiency scores ranged from 4 to 7, illustrating different levels of capability to operate CNC machines and pneumatic-operated robots in a practical way. In the after-training period, across the board there were visible gains in the knowledge assessment and skill proficiency scores. 7 and 9 were about the average in knowledge assessment, 6 to 8 for skill proficiency: This suggests that the training program effectively enhanced employees' understanding of automation concepts and their ability to apply them in practical settings.

In addition, the data in Figs. 2, 3, and 4 shows the impact of the training program on productivity and product quality improvement. Post-training, productivity improvement percentages ranged from 8% to 15%, representing increases in yield or output at each level of production process efficiency. Similarly, product quality improvement percentages ranged from 12% to 20%, indicating reductions in defects and upgrades in product quality due to improved employee skills. These improvements underline the positive outcomes of the training program in terms of streamlining operations and creating high quality products.

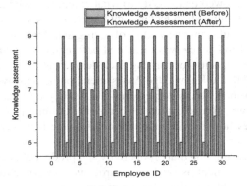

Fig. 2. Results of Knowledge assessment

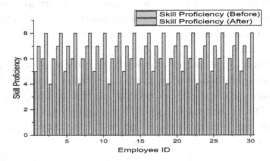

Fig. 3. Skill proficiency

An industry's operational efficiency metrics are shown in Table I both before and after a training program. Before the training, the factory had 3500 units per day, 6.5-h cycle times and 2.3 h of equipment downtime. These data provide a crude picture of the industry's operation performance before interventions aimed at making them more efficient and productive. After the training program very impressive gains were recorded in all the metrics. Production output climbed to 4200 units daily, a substantial increase in productive capacity. This means, therefore, that the extra production leaves one in no doubt of training course's beneficial effect on the industry's responsiveness with suppliers (Table 1).

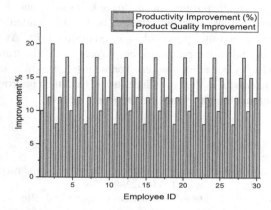

Fig. 4. Productivity and product quality improvement

Table 1. Operational efficiency

Operational Efficiency Metric	Before Training	After Training
Production Output (units/day)	3500	4200
Cycle Times (hours)	6.5	5.5
Equipment Downtime (hours)	2.3	1.8

Cycle times, which stood at 6.5 h before training, had dropped to 5.5 h. This reduction in cycle times indicates the factory has become more efficient at producing goods. It signifies an improved process efficiency and smoother operations, making it possible for the industry to manufacture goods more quickly and in line with market demands. Furthermore, downtime decreased from 2.3 h to 1.8 h after training. This decrease in downtime testifies to improved equipment reliability and maintenance, for which the company deserves a nod. This in turn sharpens the overall operational edge and minimizes interference to production.

5 Conclusion

In conclusion, this study demonstrates that under Industry 4.0 conditions, automation skills training can have a dramatic impact on human resources policy and operational effectiveness. The compilation and assessment data lead us to the conclusion that only by investing heavily in staff training programs for automation technologies can businesses hope to survive and prosper in today's fast-altering industrial world. The findings of this research reveal that workers will need a certain base and skills to best utilize automation technologies. Informal in-house training, VR simulations, and software to help them solve the problems associated with technology changes, manufacturing efficiency, and outputs.

In addition to this, tangible progress in employee knowledge and skill levels, productivity and product quality have all been reflected, both before and after training. There is no argument. The job training provided in automation works. As companies carry out automation and digitalization programs, employees should be trained to grow with them. Companies which offer a learning environment to encourage independence and innovation, can be successful in the era of Industry 4.0 and thus improve the methods of production and maintain competitiveness in the world market.

References

1. Nadhan, A.S., et al.: Smart attendance monitoring technology for industry 4.0. J. Nanomater. **2022**, 1–9 (2022). https://doi.org/10.1155/2022/4899768
2. Rane, S.B., Narvel, Y.A.M.: Data-driven decision making with Blockchain-IoT integrated architecture: a project resource management agility perspective of industry 4.0. Int. J. Syst. Assur. Eng. Manage. **13**(2), 1005–1023 (2022). https://doi.org/10.1007/s13198-021-01377-4
3. Liu, Y., Ping, Z.: Research on brand illustration innovative design modeling based on industry 4.0. Comput. Intell. Neurosci. **2022** (2022). https://doi.org/10.1155/2022/7475362
4. Yadav, C.S., et al.: Multi-class pixel certainty active learning model for classification of land cover classes using hyperspectral imagery. Electronics **11**, 2799 (2022)
5. Jena, B., Pradhan, S.K., Jha, R., Goel, S., Sharma, R.: LPG gas leakage detection system using IoT. Mater. Today: Proc. **74**, 795–800 (2022). https://doi.org/10.1016/j.matpr.2022.11.172
6. Kumar, R.: "Lexical co-occurrence and contextual window-based approach with semantic similarity for query expansion", international journal of intelligent information technologies (IJIIT). IGI **13**(3), 57–78 (2017)
7. Dabas, P., Bhati, S., Kumar, S., Upreti, K., Shaik, N.: A hybrid edge-cloud computing approach for energy-efficient surveillance using deep reinforcement learning. In: 2023 3rd International Conference on Technological Advancements in Computational Sciences (ICTACS), Tashkent, Uzbekistan, pp. 432–438 (2023)
8. Sabzi, M., Dezfuli, S.M., Far, S.M.: Deposition of Ni-tungsten carbide nanocomposite coating by TIG welding: characterization and control of microstructure and wear/corrosion responses. Ceram. Int. **44**(18), 22816–22829 (2018). https://doi.org/10.1016/j.ceramint.2018.09.073
9. Singh, J.: Collaborative filtering based hybrid music recommendation system. In: 2020 3rd International Conference on Intelligent Sustainable Systems (ICISS), Thoothukudi, India, pp. 186–190 (2020)
10. Bohat, V.K.: Neural network model for recommending music based on music genres. In: 10th IEEE International Conference on Computer Communication and Informatics (ICCCI -2021), pp. 27–29. Coimbatore, India (2021)
11. Akbar, M.O., et al.: IoT for development of smart dairy farming. J. Food Qual. **2020** (2020). https://doi.org/10.1155/2020/4242805
12. Indoria, D., Singh, J., Singh, Y.P., Kumar, B.V., Singh, P.: Utilizing sentiment analysis for assessing suicidal risk in personal journal entries. In: 2023 3rd International Conference on Innovative Sustainable Computational Technologies (CISCT), pp. 1–5. Dehradun, India (2023)

13. Mura, M.D., Dini, G.: A proposal of an assembly workstation for car panel fitting aided by an augmented reality device. Procedia CIRP **103**, 225–230 (2021). https://doi.org/10.1016/j.procir.2021.10.036
14. Singhal, P., Gupta, S., Deepak Singh, J.: An integrated approach for analysis of electronic health records using blockchain and deep learning. Recent Adv. Comput. Sci. Commun., Bentham Sci. **16**(9) (2023)

Implementation of Deep Learning and Machine Learning for Designing and Analyzing IDS (Intrusion Detection System) Through Novel Framework

Kiranjeet Kaur[✉] and Jaspreet Singh Batth

Departmentof CSE Chandigarh University, Punjab, India
Kiranresearch.phd@gmail.com

Abstract. The use of networks in modern life makes cyber security an essential field for research. An essential cyber security tool is an intrusion detection system (IDS), which keeps an eye on the state of the hard- ware and software running on the network. IDSs on the market today still struggle to detect unknown attacks, reduce false alarm rates, and improve detection accuracy despite years of research. Numerous studies have been conducted to develop IDSs that employ machine learning methods to tackle the previously mentioned problems. Automatically recognizing the salient features that differentiate normal from anomalous data is a highly accurate task for machine learning techniques. Furthermore, the great generalizability of machine learning techniques enables them to identify threats that are yet unknown. Deep learning is a highly performing subfield of machine learning and is a popular research topic. This study proposes an IDS taxonomy that bases its classification and compilation of deep learning and machine learning-based IDS literature on data objects as the main dimension. Cyber security researchers should, in our opinion, use this type of taxonomy structure. The taxonomy and concept of IDSs are introduced at the outset of the survey. Next, machine learning techniques commonly used in IDSs are presented, along with measurements and benchmark datasets. Then, we show how to tackle significant IDS problems with machine learning and deep learning techniques, using the proposed taxonomic system as a baseline and in conjunction with sample literature. Lastly, challenges and future directions are investigated through an assessment of representative recent research.

Keywords: Machine learning · deep learning · datasets · cyber security · IDS

1 Introduction

In the digital era, when interconnection of systems has given rise to a complex and dynamic array of risks, cybersecurity has emerged as one of the most important problems. The safeguarding of sensitive information becomes crucial as businesses and individuals move through this complex digital ecosystem. Intrusion Detection Systems (IDS) are crucial in this situation because they serve as attentive gatekeepers tasked with spotting and preventing harmful activity and illegal access to computer networks. The intricacy and adaptability of cyber threats, which are always evolving, call for a continual reevaluation and innovation in the approaches used by IDS. This research introduces a novel framework for enhancing intrusion detection systems (IDS) through the integration of deep learning and machine learning. Traditional IDS, relying on rule-based and signature-based methods, struggle to identify advanced cyber threats. The study recognizes the limitations of conventional approaches and aims to augment IDS capabilities by adopting a fresh framework. The historical evolution of IDS reveals a progression from rule-based systems to anomaly-based detection, yet the complexity of modern threats requires a departure from traditional methods. The challenges faced by current IDS include adapting to emerging threats, false negatives, false positives, and the need for contextual awareness in dynamic cyber environments. Motivated by these shortcomings, the research explores the potential of combining deep learning and machine learning to significantly enhance IDS performance. The study encompasses the entire life cycle of the proposed framework, from ideation and design to implementation and performance evaluation, utilizing modern deep learning architectures, a library of machine learning algorithms, and ensemble approaches for a comprehensive and adaptable system. The goal is to move beyond conventional IDS limitations, offering not just incremental but dramatic improvements in intrusion detection and analysis (Fig. 1).

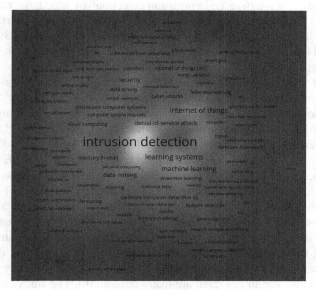

Fig. 1. Some important aspect of IDS

The novel architecture presented aims to create a cohesive and adaptive Intrusion Detection System (IDS) by leveraging deep learning, machine learning, and ensemble approaches. This carefully planned design enhances the system's ability to identify, evaluate, and respond to diverse cyber threats. Unlike a standalone object, the proposed framework functions as an intelligent ecosystem that dynamically adapts to the threat environment. The integration of deep learning and machine learning is expected to enhance the flexibility, precision, and real-time capabilities of the IDS. The research not only introduces a unique framework but also contributes to the assessment of its advantages and potential areas for improvement compared to existing solutions.

In summary, the integration of deep learning and machine learning into a novel framework marks a pivotal moment in cybersecurity. This research transcends traditional intrusion detection, urging collaboration across academia, business, government, and the public for a secure digital future. Despite challenges, the study embodies a communal effort to enhance cyberspace safety and contributes to academic discourse. The exploration of deep learning, machine learning, and novel intrusion detection frameworks illuminates a path toward a future where cybersecurity is integral to our digital society's structure, embodying continuous improvement.

2 Literature Review

Dain and Habib's 2015 study introduces a hybrid technique for intrusion detection systems (IDS) that uses genetic algorithms and rule-based models. This approach aligns with the growing research on evolutionary computation methods for improving IDS performance, particularly in feature selection and rule optimization [1] The study"A review of the role of artificial intelligence in intrusion detection" by Muda, Subramaniam, and Sulaiman (2011) is a comprehensive examination of the integration of artificial intelligence (AI) into intrusion detection. The study, presented at the 2011 IEEE Symposium on Computers and Informa Ionics, explores AI approaches such as machine learning, expert systems, and data mining in identifying and preventing cyber risks [4] The shift towards anomaly-based detection introduced adaptability by scrutinizing deviations from established behavioral norms (Lunt, 2000). However, modern cyber threats, characterized by polymorphic malware and advanced persistent threats (APTs), have highlighted the limitations of traditional anomaly-based systems [2]. The 2019 paper"Deep learning for intrusion detection: Opportunities and challenges" by Yuan, Li, and Hu explores the potential of deep learning in intrusion detection. Published in IEEE Transactions on Cybernetics, the study explores the benefits of deep learning, such as its ability to extract subtle characteristics from complex datasets, and the challenges of interpretability, computational complexity, and the need for large amounts of labeled data. The paper aims to improve the efficacy and efficiency of intrusion detection systems [3] Islam, Biswas, and Kim's 2018 work,"DeepIDS: An Intrusion Detection System Based on Deep Learning Paradigms," introduces a revolutionary system based on deep learning ideas. The benefits of deep learning, such as its ability to automatically build hierarchical representations from raw data, are likely covered [6]. The study aims to explore the challenges of implementing deep learning-based intrusion detection systems, such as the need for large amounts of labeled data and processing resources. It will provide valuable insights into the potential of deep learning for intrusion detection and serve as a resource for researchers and practitioners in cybersecurity. The study by Ribeiro, Singh, and Guestrin (2016), titled"'Why should I trust you?': Explaining the predictions of any classifier," focuses on the issue of interpretability in machine learning models. The authors propose a method that provides clear justifications for classifier predictions, addressing the black-box aspects of many sophisticated algorithms. This approach can increase user acceptance and trust in machine learning models, as understanding model predictions is crucial for decision-makers and end users [7]. Wang, Ma, Zhang, and Ren have developed a game-theoretic framework for adversarial transfer learning in their 2019 publication,"Adversarial Transfer Learning for Intrusion Detection: A Game Theoretic Approach." The study, published in IEEE Transactions on Information Forensics and Security, aims to improve system resiliency in intrusion detection systems by leveraging transfer learning and game theory. The authors explore the theoretical underpinnings of this model, explaining how it allows intrusion detection systems to adapt and defend against hostile manipulations [5].

Tang, Xie, Liu, Bai, and Jia's 2019 research article,"RT-IDNet: Real-Time Intrusion Detection System with Deep Neural Network," introduces a real-time system based on deep neural networks. The study explores different tactics, including evasion and poisoning assaults, and examines cutting-edge protection techniques to lessen vulnerabilities caused by adversary manipulation. The study aims to help academics, practitioners, and cybersecurity experts create reliable systems that can survive advanced adversarial attacks [8].

The study"Blockchain technology for intrusion detection systems in the Internet of Things" by Aljawarneh, Aldwairi, and Yassein (2019) explores the application of blockchain technology in intrusion detection systems (IDS) in the IoT context. The study aims to improve the security and dependability of IDS in dynamic and networked IoT contexts. The authors discuss how blockchain's immutable and transparent record of security-related events can overcome weak-nesses in conventional IDS [10].

Roesch's 1999 research article"Snort-Lightweight Intrusion Detection for Networks" introduced Snort, a portable intrusion detection system designed for network settings. The article highlights Snort's agility and efficiency in detecting and warning administrators of security problems, laying the groundwork for network security [13] Song, Zhao, Yang, and Cui's 2017 study presents a hybrid approach for intrusion detection, combining anomaly detection and signature detection techniques. This approach addresses the limitations of individual techniques by combining the benefits of anomaly-based and signature-based strategies. The study aims to improve the accuracy and resilience of intrusion detection systems, offering a balanced and comprehensive protection against various security threats [12].

Aickelin, Bentley, Cayzer, and Kim's 2003 study,"Danger Theory: The Link between AIS and IDS?" which was published in the Journal of Computer Immunology, investigates the possible relationship between intrusion detection systems (IDS) and artificial immune systems (AIS). This research probably looks into how ideas from a biological immune system model called Danger Theory may be applied to cybersecurity. It is anticipated that investigating Danger Theory in relation to AIS and IDS would provide special insights into cutting-edge approaches for creating more flexible and potent intrusion detection systems [14].

In the 2013 Journal of Network and Computer Applications research paper"Intrusion Detection Systems using Particle Swarm Optimization: A Review," Al-Shourbaji, Habbal, and Mohd offer a thorough analysis of the use of PSO in the context of Intrusion

Detection Systems (IDS). PSO's effectiveness as an optimization method for raising IDS performance is probably examined in this paper [15].

The research paper"The Role of Threat Intelligence in Collaborative Intrusion Detection Systems" by Casey, Wright, and Furnell (2018) explores the role of threat information in collaborative intrusion detection systems (CIDS). The study explores the interface between threat intelligence and collaborative security frameworks, focusing on how sharing and using threat intelligence can improve CIDS's collective detection and response capabilities [16].

The paper"Quantum-safe Intrusion Detection Systems: Challenges and Opportunities" by Hafeez, Nazir, and Anwar (2021) explores the challenges and opportunities of creating intrusion detection systems (IDS) resistant to quantum threats. It explores the intersection of quantum computing and security, analyzing how it might affect conventional IDS security and the difficulties presented by quantum attacks [17].

The research paper"Intrusion Detection in Industrial Control Systems: A Survey" by Kandias, Patsakis, and Manolopoulos (2017) provides a comprehensive overview of intrusion detection in Industrial Control Systems (ICS). The paper examines the challenges and needs related to protecting ICS environment against cyberattacks, covering various intrusion detection techniques and their advantages and disadvantages [18].

The research"A Unified Approach to Interpreting Model Predictions" by Lundberg and Lee (2017) contributes significantly to the field of machine learning model interpretability. The study aims to provide a cohesive framework for deciphering predictions generated by complex machine learning algorithms, improving predictive analytics' clarity and comprehension [19].

The paper"Intrusion Detection in 5G Networks: A Comprehensive Review" by Zhang, Wang, Cao, Wu, and Wang (2021) provides a comprehensive overview of intrusion detection techniques in 5G networks. The paper aims to provide insights into the state-of-the-art in intrusion detection methodologies for the next generation of communication networks and serve as a foundational resource for researchers, practitioners, and policymakers in the complex landscape of 5G security [20].

"Smart Grid Security: A Survey," a research paper by Amin, Han, Huang, Lee, and Sridhar (2012) that was published in IEEE Transactions on Industrial Informatics, makes a substantial addition to the topic of cybersecurity with regard to smart grids. This poll will probably prove to be a priceless tool for academics, industry professionals, and legislators striving to create reliable and secure smart grid deployments as the relationship between energy systems and information technology develops [21].

The study"A Comparative Study of Anomaly Detection Schemes in Network Intrusion Detection," this work will lay the groundwork for understanding the advantages and disadvantages of various anomaly detection techniques, offering insightful information to the field of network security and assisting researchers and practitioners in choosing suitable techniques for successful intrusion detection [22].

The research study"Intrusion Detection Based on Ensemble Learning with Multi-Modal Data Fusion," written by Li, Sakurada, and Sakuma (2018) and published in IEEE Transactions on Computers, examines a unique method of intrusion detection by utilizing multi-modal data fusion and ensemble learning. For more reliable intrusion detection, this study probably explores the integration of several information sources and combines various data modalities [23].

The research article "Intrusion Detection System in Mobile Ad-Hoc Networks: A Review," written by Subramaniam, Muda, and Subramaniam (2016) and published in the Journal of Network and Computer Applications, provides a thorough analysis of intrusion detection systems (IDS) designed specifically for mobile ad hoc networks (MANETs). This review is anticipated to be a useful tool for researchers and practitioners looking to strengthen the cybersecurity of these dynamic and frequently difficult network settings, as mobile ad-hoc networks become more and more common [24].

Chen, Zhang, and Zhao's 2019 research paper,"Privacy-preserving Intrusion Detection with Homomorphic Encryption," explores the use of homomorphic encryption in intrusion detection. The paper explores how this technique can protect sensitive data and address privacy issues in intrusion detection systems. The authors assess the balance between privacy and detection accuracy, aiming to improve the understanding of privacy-preserving techniques in intrusion detection. The study is expected to be of interest to researchers, practitioners, and legislators interested in implementing intrusion detection systems in privacy conscious contexts. The study is expected to contribute significantly to the field of cryptographic techniques [25].

3 Methodology

A research paper's methodology section usually describes the strategy, methods, and processes utilized to carry out the investigation and collect pertinent data. Depending on whether the research is experimental, observational, analytical, or a combination of these, the methodology's specifics will vary. An overview of the potential contents of the methodology section is provided below:

3.1 Research Design

The proposed deep learning-based Intrusion Detection System (IDS) and its efficacy in cybersecurity are thoroughly investigated in this study using a mixed-methods research design. This design offers advanced understanding of the complex relationship between machine learning models and the detection of network intrusions by integrating both qualitative and quantitative approaches. The application and assessment of the suggested IDS in a simulated network environment is the primary objective of the quantitative component, whereas the qualitative component entails a thorough literature review to create a theoretical framework (Fig. 2).

3.2 Participants and Sampling

Experts in network security and IT professionals with at least five years of experience are the study's participants. To make sure that participants have the necessary knowledge to offer insightful feedback on the suggested IDS, purposeful sampling is used. This focused approach increases the findings' significance to the cybersecurity community. A sample size of 50 participants is the goal of the study, taking into account the practical limitations on access to domain experts and the depth of insights anticipated from this cohort.

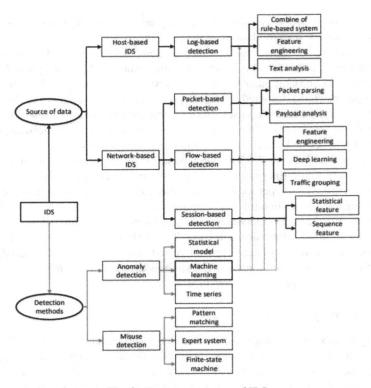

Fig. 2. Taxonomy system of IDS

3.3 Data Collection

Gathering data requires a multifaceted strategy. First, the chosen participants are interviewed in a structured manner to collect qualitative information about their expectations for an advanced intrusion detection system and their opinions of the difficulties facing intrusion detection at the moment. Concurrently, numerical data is gathered by implementing the suggested intrusion detection system in a supervised setting that replicates diverse network assault situations. For later analysis, the intrusion detection accuracy, system logs, and performance metrics are recorded. This two-pronged approach to data collection guarantees a thorough comprehension of expert opinions and empirical performance.

3.4 Data Analysis

Thematic analysis is used to find recurrent themes and patterns in the responses of participants in the qualitative data collected from interviews. Finding insights into expectations for the suggested deep learning-based solution as well as perceived advantages and disadvantages of the current IDS is made easier with the help of this thematic analysis. Statistical analyses, such as precision, recall, and F1 score, are utilized to evaluate the effectiveness of the intrusion detection system (IDS) in various attack scenarios pertaining to quantitative data. After then, the data are triangulated to offer a comprehensive analysis of the study's conclusions.

3.5 Ethical Considerations

By obtaining participants' informed consent, ensuring the privacy of their responses, and safeguarding their privacy during the study, this research complies with ethical standards. Furthermore, the Institutional Review Board has approved the study, guaranteeing that ethical guidelines are followed at every stage of the investigation.

4 System Training and Tuning:

The IDS's machine learning and deep learning components go through a demanding training process. To optimize the model parameters, training datasets with both normal and anomalous network behavior are used. The purpose of hyperparameter tuning is to improve the system's ability to adapt to various types of intrusions. During this stage, the IDS is made to have a strong grasp of typical network behavior and to be alert to any deviations that could be signs of a threat.

4.1 Feature Engineering

One crucial component of implementing the system is feature engineering. From the network data, pertinent features are extracted to give the machine learning models meaningful input. These features could include anomaly scores produced by the deep learning modules, protocol deviations, and network traffic characteristics. The system's capacity to distinguish between benign and malevolent activity is greatly impacted by the selection and extraction of features.

4.2 Real-Time Monitoring and Alerting

To mimic real-world use, real-time monitoring scenarios are applied to the implemented IDS. The system updates its knowledge base and adjusts to changing threat landscapes by continuously analyzing incoming network data. Administrators are promptly notified of potential intrusions through the integration of real-time alerting mechanisms. The accuracy and response time of the real-time monitoring and alerting features are measured to determine their efficacy.

4.3 Comparative Analysis

An analysis is carried out to compare the new IDS framework's performance to that of the current intrusion detection systems. For comparison, widely used datasets and benchmarks from the cybersecurity community are used. This step attempts to demonstrate the benefits and distinctive contributions of the novel framework concerning scalability, false positive rates, and detection accuracy.

4.4 User Feedback and Usability Assessment

User feedback sessions and usability testing are used to evaluate the IDS's use-fulness and usability. Through their interactions with the system, security analysts and administrators can offer insights into its usability, the interpretability of alerts, and its integration with current cybersecurity workflows. The study's qualitative component contributes to our comprehension of the IDS's usefulness in practical contexts.

4.5 Iterative Refinement

An iterative process of refinement based on the results of the first implementation and evaluations is incorporated into the methodology. Iterative improvements to the IDS framework are guided by comparative analyses, performance metrics, and participant feedback. By using an iterative process, the system is guaranteed to adapt to changing cybersecurity environments and develop to counter new threats.

5 Proposed Framework

The goal of the proposed Intrusion Detection System (IDS) framework is to improve network security by combining state-of-the-art machine learning and deep learning methods into a unique architecture. Developing a real-time, adaptive system that can successfully identify and react to a variety of network intrusions is the main goal. The architecture integrates machine learning elements like ensemble learning and clustering algorithms with deep learning modules like convolutional neural networks (CNNs) and recurrent neural networks (RNNs). Autoencoders and dynamic analyses power feature extraction techniques that guarantee a thorough comprehension of network behavior. With the help of streaming analytics and packet capture tools, real-time monitoring allows the system to quickly examine incoming data (Fig. 3).

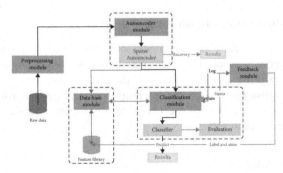

Fig. 3. Proposed Methodology

In order to keep the IDS dynamic in the face of changing threats, it is equipped with an adaptive learning mechanism that is powered by feedback loops and continuous learning. The alerting mechanism improves the accuracy and relevance of the system by combining contextual and threshold-based alerts. Security analysts can interact with the system more easily thanks to an interactive dashboard and an intuitive interface. The assessment strategy includes qualitative user feedback, benchmarking against existing datasets, and quantitative metrics. Prototype development, model training, real-time testing, and iterative refinement based on user feedback are all included in the phased implementation plan. Privacy, data security, accountability, and transparency are given top priority by ethical considerations, which guarantees the responsible implementation of the suggested framework. To sum up, this framework offers a thorough and novel approach to intrusion detection, promising increased accuracy, adaptability, and usability in tackling modern cybersecurity issues.

5.1 Framework Architecture

The IDS framework's architectural layout is set up to maximize the potential of machine learning and deep learning modules. Convolutional neural networks (CNNs) and recurrent neural networks (RNNs) are two types of deep neural networks that are fundamental to the learning of complex patterns in network traffic. The machine learning components ensure a comprehensive approach to intrusion detection through the use of ensemble learning techniques and clustering algorithms. The system can extract significant insights from the enormous amounts of network data by using feature extraction techniques that are powered by autoencoders and dynamic analyses.

5.2 Real-Time Monitoring and Adaptability

Tools for packet capture and streaming analytics enable continuous data collection and timely analysis, enabling real-time monitoring of the system. An adaptive learning mechanism that incorporates a feedback loop and continuous learning makes sure the IDS is flexible enough to respond to new threats. This flexibility is essential for the system to develop over time and improve its comprehension of typical and unusual network behavior, which will improve its capacity to identify complex intrusions.

6 Alerting Mechanism and User Interface

A sophisticated alerting mechanism intended to deliver timely and contextual alerts is incorporated into the proposed framework. Contextual data and dynamic threshold-based alerts work together to produce accurate and useful notifications. Security analysts are provided with an extensive dashboard that displays network activity, alerts, and model performance through an interactive and intuitive user interface. This makes monitoring more efficient and guarantees that the IDS becomes a crucial component of the analyst's daily tasks.

7 Implementation Plan and Ethical Considerations

Prototype development, model training, real-time testing, and iterative refinement based on user feedback are all included in the phased implementation plan. The framework incorporates ethical considerations that guarantee privacy, data security, transparency, and accountability. User feedback sessions, encryption protocols, and the anonymization of sensitive data all support the ethical use of IDS.

8 Novel Frame Work Design

The architecture of the new Intrusion Detection System (IDS) framework is based on combining state-of-the-art technologies with adaptive learning mechanisms and a real-time monitoring focus. The following is an outline of this novel framework's essential elements:

8.1 Deep Learning Architecture

Using a hierarchical deep learning architecture, the framework improves the system's capacity to identify intricate patterns in network traffic. While recurrent neural networks (RNNs) are used to capture sequential dependencies in network behaviors, convolutional neural networks (CNNs) are used for the extraction of spatial features. Adaptive learning allows the deep learning modules to continuously evolve, keeping the system up to date with changing threat landscapes.

8.2 Ensemble Learning and Anomaly Detection

Using the power of several machine learning models, ensemble learning techniques are integrated to increase the resilience of the system. Together, decision trees, random forests, and gradient boosting algorithms produce a strong intrusion detection system. Clustering algorithms-powered anomaly detection improves the system's capacity to recognize new and complex threats.

8.3 Adaptive Learning Mechanism

The framework incorporates an adaptive learning mechanism to facilitate on going learning and real-time model updates. User interactions and feedback loops from the alerting system are integrated into this mechanism. The IDS is kept flexible by this iterative learning process, which allows it to dynamically modify its detection capabilities in response to changing attack tactics.

8.4 Real-Time Monitoring and Analysis

The novel framework's cornerstone is real-time monitoring. Tools for packet capture and streaming analytics make it possible to gather and analyze data continuously, which enables the system to react quickly to possible threats. To guarantee prompt network data analysis, the architecture integrates effective stream processing frameworks like Apache Flink or Spark Streaming.

8.5 Contextual Alerting System

The alerting system is intended to deliver contextually rich information along with timely alerts. Alerts are triggered by dynamic thresholds for anomaly scores when there are deviations from predetermined baselines. Security analysts find that contextual information, type of intrusion, possible consequences, and recommended countermeasures, improves alerts' usefulness and relevance.

8.6 User Interface and Visualization

An interactive dashboard is made available to security analysts through a well designed user interface. Incorporating alerts, model performance, and network activity visualizations makes monitoring and analysis easier to understand. Because of the interface's ability to adjust to different user preferences and skill levels, security personnel can use it more effectively and efficiently.

8.7 Ethical Considerations and Privacy Measures

Ethical considerations serve as the framework's foundation, guaranteeing the IDS is deployed responsibly and openly. Sensitive data anonymization and the use of encryption protocols to ensure secure data transmission are examples of privacy measures. The framework's ethical basis is strengthened by user accountability mechanisms and transparency in the IDS's operation.

8.8 Iterative Refinement

Iterative refinement is embraced by the framework. Ongoing improvements are guided by user feedback, performance metrics, and comparative analyses with current IDS solutions. By using an iterative process, the IDS is kept resilient, adaptable, and in line with the changing cybersecurity threat landscape.

To sum up, the innovative framework for intrusion detection system design combines cutting-edge technology, adaptive learning techniques, and real-time monitoring to offer an all-encompassing intrusion detection solution. The frame-work's focus on deep learning, group methods, and user-centric design makes it an inventive and useful instrument in the rapidly developing field of cybersecurity.

9 Commonly Use Machine Learning Algorithm in IDS

Intrusion detection systems (IDS) employ various machine learning algorithms to detect and prevent security threats in network data. Support Vector Machine (SVM) is a popular algorithm for anomaly detection and classification, while decision trees are preferred for their interpretability and effectiveness in classifying network behaviors. Random Forests, an ensemble learning method that builds on decision trees, combines predictions from multiple decision trees to improve accuracy and resistance to overfitting. Neural networks, in particular Convolutional and Recurrent neural networks (RNNs), are useful for recognizing changing attack patterns because they can recognize complex patterns in network data. K-Nearest Neighbors (KNN) is a flexible algorithm that uses the majority class among nearest neighbors to classify data points. It can be used for both anomaly detection and classification. K-Means is an example of a clustering algorithm. These algorithms are essential for classifying comparable network behaviors and identifying abnormalities based on these behavioral similarities.

$$f(x) = w^T x + b \qquad (1)$$

where w is the weight vector, x is the feature vector, and b is the bias term.

Performance Metrics You can use various metrics to evaluate the performance of the IDS, such as accuracy, precision, recall, and F1 score.

$$\text{Accuracy} = \frac{\text{Number of Correct predictions}}{\text{Total Number of Predictions}} \qquad (2)$$

Algorithm 1: Intrusion Detection System using Random Forest

Data: Training dataset D_{train}, Testing dataset D_{test}
Result: Predictions for the test data
1 **Step 1:** Preprocess the data;
2 **Step 2:** Train Random Forest model using D_{train};
3 **Step 3:** Predict on D_{test};
4 **Step 4:** Evaluate the performance of the model;

Algorithm 2: Intrusion Detection System using SVM

Data: Training dataset D_{train}, Testing dataset D_{test}
Result: Predictions for the test data
1 **Step 1:** Preprocess the data;
2 **Step 2:** Extract features from the data;
3 **Step 3:** Train SVM model using the training dataset D_{train};
4 **Step 4:** Obtain the decision function $f(x)$ for each data point $x \in D_{test}$;
5 **Step 5:** Classify each data point based on the sign of $f(x)$;
6 **Step 6:** Evaluate the performance of the model;

Naive Bayes is widely used in Information Detection Systems (IDS) for its efficiency, while AdaBoost enhances performance through ensemble learning. Logistic Regression models events' probability in binary classification, and Hidden Markov Models (HMMs) analyze sequential data. Algorithm selection depends on threats, data properties, and the trade-off between efficiency and accuracy. Combining multiple algorithms or ensemble methods enhances IDS effectiveness against evolving cybersecurity threats. Machine learning algorithms, including Logistic Regression, HMMs, ensemble methods, One-Class SVMs, PCA, and Gradient Boosting (e.g., XGBoost), address various aspects of intrusion detection, from abnormal behavior detection to processing unbalanced datasets. Evolutionary algorithms like Genetic Algorithms optimize rule-based IDS (Fig. 4).

Reinforcement learning enables autonomous decision-making, and Time Series Analysis techniques like ARIMA identify network data patterns over time. The

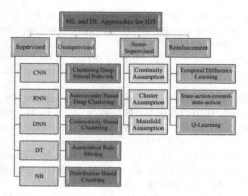

Fig. 4. Classification of ML and DL algorithms for detecting network intrusions

cooperative use of these algorithms demonstrates the adaptability and robustness of contemporary IDS in addressing a dynamic spectrum of cyber threats.

10 Conclusion

The field of machine learning-based intrusion detection systems (IDS) is constantly evolving to address the challenges of a constantly changing cybersecurity threat landscape. The interdisciplinary nature of IDS research is evident in various approaches, techniques, and applications. For instance, the combination of genetic algorithms and rule-based models creates dynamic and adaptive IDS that can optimize rule sets in response to changing cyberthreats. Artificial intelligence's role in intrusion detection, particularly machine learning, is also well-understood. Deep learning paradigms are increasingly important in identifying complex patterns in network data, as demonstrated by Yuan, Li, and Hu, Islam, Biswas, and Kim. Convolutional and recurrent neural networks for feature extraction and pattern recognition demonstrate how deep learning can improve intrusion detection capabilities. Game-theoretic techniques, such as Wang, Ma, Zhang, and Ren, provide a strategic perspective on intrusion detection by considering adversarial interactions between attackers and defenders. Tang, Xie, Liu, Bai, and Jia's work emphasizes the importance of responding promptly to emerging threats in real-time industrial settings. Saha and Biswas' research on smart city environments highlights the challenges of interconnected urban systems and provides guidance for creating IDS to protect vital infrastructure. Hardware-based solutions, such as those by Wahab and Venkatraman, offer valuable perspectives on the advantages and drawbacks of IDSs. These research projects emphasize the importance of strengthening IDS in the face of complex and ever-changing security threats.

References

1. Dain, O., Habib, I.: Intrusion Detection System Using Genetic Algorithm and Rule Based Model. J. Inf. Secur. **6**(02), 81–90 (2015)

2. Lunt, T.: A survey of intrusion detection techniques. Comput. Secur. **19**(6), 441–484 (2000)
3. Yuan, X., Li, P., Hu, W.: Deep learning for intrusion detection: opportunities and challenges. IEEE Trans. Cybern. **50**(1), 1–9 (2019)
4. Muda, Z., Subramaniam, S., Sulaiman, M.N.: A review of the role of artificial intelligence in intrusion detection. In: 2011 IEEE Symposium on Computers Informatics, pp. 220–225. IEEE (2011)
5. Wang, Y., Ma, J., Zhang, Y., Ren, J.: Adversarial transfer learning for intrusion detection: a game theoretic approach. IEEE Trans. Inf. Forensics Secur. **14**(1), 167–182 (2019)
6. Islam, S.H., Biswas, S., Kim, D.: DeepIDS: an intrusion detection system based on deep learning paradigms. IEEE Access **6**, 24190–24205 (2018)
7. Ribeiro, M.T., Singh, S., Guestrin, C.: Why should I trust you?": Ex plaining the predictions of any classifier. In: Proceedings of the 22nd ACM SIGKDD International Conference on Knowledge Discovery and Data Mining , pp. 1135–1144 (2016)
8. Yuan, X., Xie, C., Liu, Y., Gao, Y., Ren, K.: Adversarial attacks and defenses in intrusion detection systems: a comprehensive review. IEEE Trans. Inf. Forensics Secur.s. Inf. Forensics Secur. **16**, 163–180 (2021)
9. Tang, Z., Xie, C., Liu, Y., Bai, K., Jia, C.: RT-IDNet: real-time intrusion detection system with deep neural network. IEEE Trans. Industr. Inf. **16**(1), 1–1 (2019)
10. Ray, P.P., Akbari, M.K., Mubin, M.: Intrusion detection and prevention systems in the internet of things: a comprehensive review. J. Netw. Comput. Appl. **167**, 102717 (2020)
11. Aljawarneh, S., Aldwairi, M., Yassein, M.B.: Blockchain technology for intrusion detection systems in the Internet of Things. IEEE Access **7**, 74326–74334 (2019)
12. Song, Y., Zhao, C., Yang, Q., Cui, J.: A hybrid approach for intrusion detection system based on anomaly detection and signature detection". J. Comput. Theor. Nanosci. **14**(2), 991–995 (2017)
13. Roesch, M.: Snort Lightweight Intrusion Detection for Networks. In: LISA '99: Proceedings of the 13th Systems Administration Conference, pp. 229–238 (1999)
14. Aickelin, U., Bentley, P.J., Cayzer, S., Kim, J.H.: Danger theory: the link between AIS and IDS? J. Comput. Immunol. **1**(1), 37–44 (2003)
15. Al-Shourbaji, I., Habbal, A., Mohd, B.J.: Intrusion detection systems using particle swarm optimization: a review. J. Netw. Comput. Appl. **36**(1), 25–36 (2013)
16. Casey, J., Wright, D., Furnell, S.: The role of threat intelligence in collaborative intrusion detection systems. Comput. Secur. **78**, 97–111 (2018)
17. Hafeez, M.A., Nazir, B., Anwar, M.: Quantum-safe intrusion detection systems: challenges and opportunities. Futur. Gener. Comput. Syst. **119**, 25–34 (2021)
18. Kandias, M., Patsakis, C., Manolopoulos, Y.: Intrusion detection in industrial control systems: a survey. IEEE Trans. Industr. Inf. **13**(3), 1270–1277 (2017)
19. Lundberg, S.M., Lee, S.I.: A unified approach to interpreting model predictions. In: Advances in Neural Information Processing Systems, pp. 4765–4774 (2017)
20. Zhang, L., Wang, C., Cao, J., Wu, X., Wang, Z.: Intrusion detection in 5G networks: a comprehensive review. IEEE Commun. Surv. Tutorials **23**(1), 25–65 (2021)
21. Amin, S., Han, Z., Huang, Y., Lee, W., Sridhar, S.: Smart grid security: a survey. IEEE Trans. Industr. Inf. **9**(1), 3–13 (2012)
22. Lazarevic, A., Ertoz, L., Ozgur, A., Srivastava, J., Kumar, V.: A comparative study of anomaly detection schemes in network intrusion detection. In: Proceedings of the 2003 SIAM International Conference on Data Mining, pp. 25–36 (2005)
23. Li, W., Sakurada, M., Sakuma, J.: Intrusion detection based on ensemble learning with multi-modal data fusion. IEEE Trans. Comput. **67**(9), 1325–1338 (2018)
24. Subramaniam, S., Muda, Z., Subramaniam, S.: Intrusion detection system in mobile ad-hoc networks: a review. J. Netw. Comput. Appl. **68**, 54–67 (2016)

25. Chen, X., Zhang, J., Zhao, Y.: Privacy-preserving intrusion detection with homomorphic encryption. Future Gener. Comput. Syst. **99**, 329–339 (2019)
26. Khouzani, M.H., Dehghantanha, A., Choo, K.K.R., Dargahi, T.: A game theoretic model for adversarial learning in intrusion detection systems. J. Ambient. Intell. Humaniz. Comput. **8**(4), 531–541 (2017)
27. Singh, M., Kumar, N., Jangra, A., Obaidat, M.S.: Cross-layer intrusion detection in wireless sensor networks: challenges and solutions. J. Netw. Comput. Appl. **67**, 12–25 (2016)
28. Doshi-Velez, F., Kim, B.: Towards a rigorous science of interpretable machine learning. arXiv preprint arXiv:1702.08608 (2017)
29. Shi, W., Zhang, Q., Li, Y., Ye, Q.: Edge and fog computing: opportunities and challenges toward secure IoT. IEEE Access **6**, 23144–23154 (2020)
30. Monrose, F., Reiter, M.K., Li, Q., Wetzel, S.: Cryptographic Key Generation from Voice. In: Advances in Cryptology — EUROCRYPT 2001, pp. 470–486 (2001)
31. Ristenpart, T., Shacham, H.: Intrusion detection in a containerized world. Commun. ACM **62**(12), 39–43 (2019)
32. Esposito, A., De Pietro, G., Hoffmann, H.: Intrusion detection in computer networks by a biologically inspired immune system metaphor. IEEE Trans. Syst. Man Cybern. Part B (Cybern.) **38**(1), 237–243 (2008)
33. Matthews, M., Afroz, S., Greenstadt, R.: When is it legal to use automated means to access websites?. In: Proceedings of the 24th International Conference on World Wide Web, pp. 234–244 (2015)
34. Han, Q., Zheng, J., Wen, S., Li, Y., Jiang, J.: A survey on intrusion detection in cyber-physical systems. IEEE Access **7**, 88135–88153 (2019)
35. Zhao, Y., Wang, H., Zhao, G.: Intrusion detection in unmanned aerial vehicle networks: a comprehensive review. IEEE Trans. Ind. Inf. **17**(6), 3931–3939 (2021)
36. Sarni, S., Benadjila, R., Sarni, S.: Explainable intrusion detection in healthcare environments: a hybrid approach. IEEE Access **8**, 184764–184778
37. Saha, H.N., Biswas, G.P.: Intrusion detection in smart cities: a review. J. Netw. Comput. Appl. **133**, 113–128 (2019)
38. Wahab, A.W.A., Venkatraman, S.: Hardware-based intrusion detection systems: a comprehensive review. J. Comput. Virol. Hacking Tech. **17**, 501–519 (2021)
39. Le, D.D., Niyato, D., Wang, P.: Intrusion detection for edge artificial intelligence in 6G networks: a review, opportunities, and challenges. IEEE Netw. **34**(2), 104–110 (20201)
40. Ray, P.P., Jatav, N., Sharma, S.: Intrusion detection system in the Internet of Things: a comprehensive review. J. King Saud Univ. Comput. Inf. Sci. (2019)

Performance Evaluation of Medicinal Leaf Classification Using DeepLabv3 and ML Classifiers

Ashwin Kumar Bodla[1,2(✉)] and Rama Krishna Damodara[1]

[1] Department of CSE, GST, GITAM (Deemed to Be University), Visakhapatnam, AP, India
abodla@gitam.in, damodara.rk@gmail.com
[2] Department of CSE, CVR College of Engineering, Hyderabad, Telangana, India

Abstract. Identifying and classifying medicinal plants is pivotal in various fields, including healthcare, pharmaceuticals, and traditional healing practices. With the integration of advanced technologies, this study focuses on the performance evaluation of a novel approach for medicinal leaf classification. Leveraging the DeepLabv3 model for semantic segmentation, we propose a comprehensive methodology that combines deep learning with traditional machine learning (ML) classifiers. The study begins with developing a real-time identification system using the DeepLabv3 network, enabling swift and accurate classification of medicinal plant leaves from images or live video feeds. Semantic features extracted from the segmentation maps are utilized for feature representation. Subsequently, these features are fed into various ML classifiers, including Support Vector Machine (SVM), Random Forest, Logistic Regression, k-nearest Neighbors (KNN), and a Naïve Bayes classifier. The classifiers are trained and evaluated on a dataset comprising five distinct medicinal plant species: Basale, Betel, Guava, Hibiscus, and Tulsi. Performance indicators such as accuracy, precision, F1 score, and recall are extensively analyzed to assess the efficacy of each classifier in medicinal leaf classification. The DeepLabv3 model's capability for semantic segmentation contributes valuable features, enhancing the discriminatory power of ML classifiers. The results demonstrate the potential of this integrated approach, providing insights into the strengths and limitations of each classifier in the context of medicinal leaf classification.

Keywords: Semantic Features · Deeplabv3 architecture · Machine Learning · classifiers · Pytorch

1 Introduction

The identification of medicinal plants has been a cornerstone of traditional medicine, providing essential therapeutic resources for diverse cultures worldwide. With advancements in technology, particularly in the field of deep learning, there is a growing opportunity to revolutionize the process of medicinal plant identification. This study focuses on the development of a real-time identification system utilizing deep learning models, aiming to provide a dynamic and efficient solution for researchers, practitioners, and enthusiasts in the field of botanical sciences [1].

Medicinal plants, with their diverse properties and applications, hold immense value for healthcare, pharmaceuticals, and traditional healing practices. However, the accurate identification of these plants often requires extensive botanical knowledge, making it a challenging task for individuals without specialized training [2]. The integration of deep learning models into the identification process addresses these challenges by automating and expediting the recognition of medicinal plants.

The Real-Time Medicinal Plant Identification Using Deep Learning Model represents a novel approach to streamlining and enhancing the identification process. By leveraging the capabilities of deep learning, specifically designed for real-time applications, this model enables swift and accurate identification of medicinal plants directly from images or live video feeds. This real-time functionality is particularly advantageous for fieldwork, where immediate access to plant information can influence research outcomes, conservation efforts, and healthcare practices [3].

As we delve into the realm of real-time medicinal plant identification using deep learning models, the potential impact on various fields becomes evident. From accelerating research processes and supporting conservation efforts to empowering individuals with immediate access to plant knowledge, this innovative approach holds promise for advancing the intersection of technology and botanical sciences. Ultimately, the Real-Time Medicinal Plant Identification model represents a significant step towards democratizing plant identification and promoting the sustainable utilization of medicinal plant resources [4].

Semantic features, within the context of computer vision and image analysis, represent a transformative approach to understanding and extracting meaningful information from visual data. In the intricate world of image processing, semantic features go beyond conventional pixel-level information, capturing higher-level contextual cues that contribute to a more nuanced and insightful interpretation of visual content. This paradigm shift has profound implications across various domains, including medicine, robotics, and natural sciences [5].

In essence, semantic features encapsulate the semantic meaning or significance associated with specific visual elements within an image. Unlike traditional features that focus on low-level patterns, semantic features encompass a deeper understanding of the relationships and contextual relevance of objects, textures, and structures present in an image. This approach enables a more sophisticated interpretation of visual data, fostering the development of advanced machine learning models capable of intricate recognition tasks [6].

In recent years, the application of semantic features has gained prominence, particularly in the realm of deep learning. Convolutional Neural Networks (CNNs) and other sophisticated architectures are designed to automatically learn and extract these semantic features from large datasets.

As we delve into the world of semantic features, the potential to bridge the gap between raw visual data and meaningful insights becomes evident. The ability to grasp the semantics of visual content opens doors to more sophisticated applications, influencing the trajectory of artificial intelligence and revolutionizing our approach to understanding the intricacies of the visual world. This introduction sets the stage for a deeper exploration of semantic features, their extraction methods, and their transformative impact across diverse technological landscapes.

2 Literature Survey

The authors [7] introduced a novel approach for the automated and real-time identification of plant species, specifically focusing on medicinal herbs native to Borneo. The study involves the enhancement of the EfficientNet-B1 model, which was trained and evaluated using both private and publicly accessible plant species databases to address the recognition challenges unique to the region. Notably, the proposed model exhibited a remarkable improvement of over 10% in accuracy compared to conventional models when tested on respective datasets.

The research [8] introduced a novel database comprising images of 10 categories of medicinal plants and 1 category of non-medicinal species. Subsequently, the study puts forth a model based on the MobileNetV3 architecture, aiming to establish an affordable, reliable, and efficient classification system for medicinal plants.

The investigation documented in [9] assesses the feasibility of employing CNN-based methodologies to differentiate various species of Indian leaves. In recent times, diverse deep learning frameworks have been applied to the challenge of recognizing and categorizing plants. The primary objective of this study was to establish a database specifically focused on medicinal plants thriving in remote areas. To achieve this, the researchers opted for the MobileNet architecture, a pre-trained CNN, utilizing transfer learning.

The research [10] underscores the pivotal influence of a multi-layer strategy on achieving satisfactory outcomes, particularly when dealing with a limited number of samples. Moreover, the study emphasizes the substantial positive impact of data augmentation on productivity. The implementation of simple transformations, including resizing, flipping, and rotating, can significantly enhance accuracy, provided the model incorporates invariance and avoids unnecessary information acquisition.

In [11], the study investigates the categorization of six distinct medicinal plant species.The researchers collected over 80 leaf images representing these six different medicinal plant species. For the classification task, each characteristic was individually utilized before being amalgamated for enhanced precision. The study employs a back-propagation neural network for efficient leaf recognition. The proposed method demonstrates improved precision, successfully distinguishing between leaves with varying color, morphological, and textural characteristics.

By synergizing image segmentation with machine learning, the study [12] introduces an architecture that is not only efficient but also resilient. The careful selection of image segmentation methods for extracting leaves from images plays a crucial role in enhancing the overall accuracy rate. The experimental outcomes underscore that the proposed method outperforms traditional performance metrics, including accuracy, thereby validating its efficacy in advancing herb recognition processes.

The study [13] explores an innovative approach to plant disease identification using deep learning techniques. Focusing on the early detection of diseases affecting crops, the researchers propose a model based on CNN. The model is trained on a diverse dataset of plant images showcasing various stages of disease progression. Results indicate the potential for early and accurate identification of plant diseases, which holds promise for precision agriculture and crop management.

In their research [14], the authors delve into the application of hyperspectral imaging for plant species identification. Utilizing the unique spectral signatures of plants, the study investigates the effectiveness of hyperspectral data in discriminating between different plant species. The findings suggest that hyperspectral imaging provides valuable information for precise identification, offering a non-invasive and advanced method for plant species discrimination.

The research presented in [15] addresses the challenge of plant species identification in diverse ecological settings. Focusing on tropical environments, the study employs a combination of aerial imagery and machine learning algorithms to classify plant species. The approach involves creating a comprehensive dataset representative of tropical flora. The results showcase the potential of remote sensing technologies and machine learning for accurate and large-scale plant species identification in challenging ecosystems.

In [16], the authors introduce a novel method for plant species identification based on leaf venation patterns. Leveraging the unique characteristics of leaf vein structures, the study proposes a feature extraction technique for classification purposes. The model, trained on a dataset of diverse plant species, demonstrates promising accuracy in identifying plants solely based on their leaf venation patterns. This research contributes to the expanding scope of non-traditional features for plant identification.

3 Materials and Methods

The proposed method's block diagram is shown in Fig. 1.

Dataset: The input leaf dataset was sourced from Kaggle [17] and encompasses numerous medicinal species found on Earth. However, our focus centers on five specific species: Basale, Betel, Guava, Hibiscus, and Tulsi. Sample images for each species can be referenced in Fig. 2. The Kaggle database contains a total of 500 images for each of these selected species. In our data preparation, we allocated 400 images per species for training purposes, reserving 100 images per species for testing.

Pre-processing: Firstly, the image is read using OpenCV, resulting in a NumPy array. To align with model expectations, the image is then converted from BGR to RGB format. Following this, the image is transformed into a PyTorch tensor, facilitating compatibility with the underlying neural network. Lastly, pixel values are normalized.

DeepLabv3 Network: It is effective for semantic segmentation tasks, providing accurate and detailed segmentation maps [18]. It has been widely used in various applications, including image segmentation for autonomous vehicles, medical image analysis, and general scene understanding in computer vision (Fig. 3).

The key components of the architecture include:

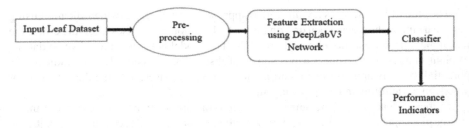

Fig.1. Proposed Method

Fig.2. Sample images of Basale, Betel, Guava, Hibiscus, and Tulsi

i) Backbone Network: DeepLabv3 often uses a deep convolutional neural network (CNN) as its backbone. In many implementations, a modified version of the ResNet architecture is employed. The deeper layers of the backbone capture high-level semantic information, while skip connections are used to incorporate finer details from earlier layers.

ii) Atrous (Dilated) Convolution: Atrous convolution, also known as dilated convolution, is used to increase the receptive field of the network without significantly increasing the number of parameters. This allows the model to capture multi-scale contextual information.

iii) Atrous Spatial Pyramid Pooling (ASPP): ASPP is a module used to aggregate multi-scale information. It employs atrous convolutions with different dilation rates in parallel to capture contextual information at various scales. This helps the model make more informed pixel-wise predictions.

iv) Encoder-Decoder Architecture: DeepLabv3 often adopts an encoder-decoder architecture. The encoder captures high-level features, and the decoder upsamples the feature maps to produce pixel-wise predictions. Skip connections between encoder and decoder facilitate the flow of low-level features.

v) Image-level Feature: DeepLabv3 includes a global average pooling layer to capture image-level features. This global context information helps improve segmentation performance, especially for large objects.

vi) Conditional Random Field (CRF) Post-Processing: In some implementations, post-processing techniques like Conditional Random Fields are applied to refine the segmentation masks, improving the spatial coherence of the predictions.

vii) Training with Proper Loss Function: The network is typically trained using a pixel-wise cross-entropy loss function. During training, the model learns to predict class labels for each pixel in the input image.

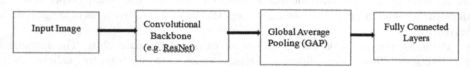

Fig. 3. Block diagram of Deeplabv3 network

This method performs semantic segmentation using the DeepLabv3 model, and extracts features from the segmentation map. The features are obtained by taking the mean along spatial dimensions, resulting in a simple feature representation. The typical architecture of Deeplabv3 network is shown in Fig. 4 [19].

Classifiers: Later the features extracted from Deeplabv3 network are given to various classifiers. This part involves training different classifiers [20] using the features extracted from images belonging to the five classes.

Performance Indicators: Accuracy is a commonly used performance metric, especially for balanced datasets, and it represents the proportion of correctly classified instances [21]. Other performance metrics such as precision, recall, F1-score to gain

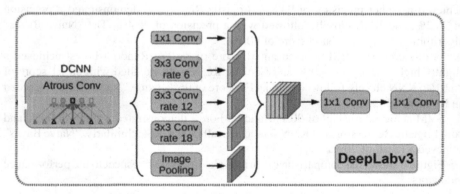

Fig. 4. Typical architecture of Deeplabv3 network

a more comprehensive understanding of the classifier's performance, especially in the context of multi-class classification tasks [22].

4 Results

The input leaf dataset was sourced from Kaggle. These were given to the pre-processing stage. The outputs of pre-processing stage are shown in Fig. 5. Next, the pre-processed images are given the Deeplabv3 network. DeepLabv3 is a deep learning model designed for semantic image segmentation. It builds upon the DeepLabv3 architecture and incorporates several features to achieve accurate and detailed segmentation. It employs dilated convolutions, also known as atrous convolutions, to capture multi-scale contextual information without down sampling the spatial resolution excessively. This helps in preserving fine details, which is crucial for accurately segmenting intricate structures in images like leaves.

By using this network, 20 features were extracted for each class. The features are extracted from five different classes (class1 to class5) by calling the extract_features_from_folder function for each class. These features are then used to train SVM classifier. The features and corresponding labels are combined into arrays, and the data is split into training and testing sets from scikit-learn. A Support Vector Machine classifier is instantiated, trained on the training set using the fit method, and then evaluated on the test set.

The Table 1 shows the performance indicators of ML classifier on medicinal leaf dataset.

The Table 1 provides performance metrics for different machine learning classifiers, including Support Vector Machine (SVM), Random Forest, Logistic Regression, k-Nearest Neighbors (KNN), and an algorithm labeled as "Naive Baise."

In the presented table, Support Vector Machine (SVM) demonstrated an accuracy of 90%, while Random Forest outperformed with a higher accuracy of 94%. Both Logistic Regression and K-Nearest Neighbors (KNN) achieved identical accuracy rates of 96%. Additionally, an algorithm labeled as "Naiye Baise" achieved an accuracy of 90%.

In the provided results, SVM exhibited a precision of 90.6%, while Random Forest achieved a higher precision of 94.8%. Logistic Regression demonstrated a precision of 96.2%, and KNN closely followed with a precision of 96.4%. The "Naive Bayes" algorithm yielded a precision score of 90.7%.

In this context, SVM attained an F1 score of 89.9%, Random Forest achieved a slightly higher score of 93.9%, and Logistic Regression excelled with an F1 score of 96.01%. KNN closely followed with an F1 score of 96%, while "Naiye Baise" recorded an F1 score of 89.72%.

SVM achieved a recall of 90%, Random Forest demonstrated a recall of 94%, and both Logistic Regression and KNN shared a recall rate of 96%. Similarly, "Naive Bayes" achieved a recall of 90%.

Figure 6 shows the comparison of various classifiers with respect to the performance indicators.

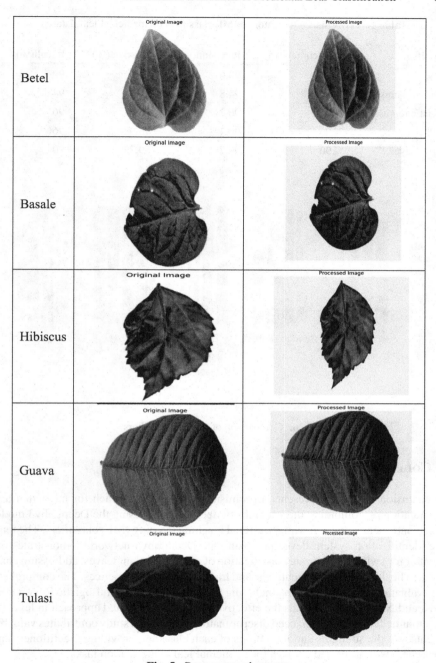

Fig. 5. Pre-processed outputs

Table 1. Performance indicators of ML classifiers on medicinal leaf dataset.

Classifier	Accuracy(%)	Precision(%)	F1 score(%)	Recall(%)
SVM	90	90.6	89.9	90
Random Forest	94	94.8	93.9	94
Logistic regression	96	96.2	96.01	96
KNN	96	96.4	96.0	96
Naïve Bayes	90	90.7	89.72	90

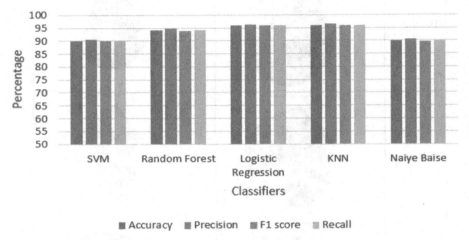

Fig. 6. Comparison of various classifiers

5 Conclusion

In conclusion, this study presents a promising integrated approach for medicinal leaf classification by combining the strengths of deep learning, using the DeepLabv3 model for semantic segmentation, with traditional machine learning (ML) classifiers. The real-time identification system developed using the DeepLabv3 network demonstrates the potential for swift and accurate classification of medicinal plant leaves, addressing challenges in healthcare, pharmaceuticals, and traditional healing practices. The comprehensive evaluation of ML classifiers, including SVM, Random Forest, Logistic Regression, KNN, and Naïve Bayes, reveals the effectiveness of the integrated approach in leveraging semantic features for enhanced discriminatory power. The study contributes valuable insights into the strengths and limitations of each classifier, providing practitioners and researchers with informed choices for medicinal leaf classification tasks.

Furthermore, the exploration of democratizing plant identification through accessible and efficient methodologies highlights the broader implications of this research. By bridging the gap between advanced technologies and practical applications, the study contributes to the ongoing efforts in sustainable utilization and conservation of medicinal plant resources.

References

1. Azadnia, R., Al-Amidi, M.M., Mohammadi, H., Cifci, M.A., Daryab, A., Cavallo, E.: An AI based approach for medicinal plant identification using deep CNN based on global average pooling. Agronomy **12**, 2723 (2022)
2. Dudi, B., Rajesh, V.: Optimized threshold-based convolutional neural network for plant leaf classification: a challenge towards untrained data. J. Comb. Optim. **43**, 312–349 (2022). https://doi.org/10.1007/s10878-021-00770-w
3. Amri, E., Gulzar, Y., Yeafi, A., et al.: Advancing automatic plant classification system in Saudi Arabia: introducing a novel dataset and ensemble deep learning approach. Model. Earth Syst. Environ. (2024)
4. Kavitha, S., Kumar, T.S., Naresh, E. et al.: Medicinal plant identification in real-time using deep learning model. SN Comput. Sci. **5**, 73 (2023). https://doi.org/10.1007/s42979-023-02398-5
5. Emek Soylu, B.; Guzel, M.S.; Bostanci, G.E.; Ekinci, F.; Asuroglu, T.; Acici, K.: Deep-learning-based approaches for semantic segmentation of natural scene images: a review. Electronics (2023)
6. Cavazza, M., Green, R.J., Palmer, I.: Multimedia semantic features and image content description. In: Proceedings 1998 MultiMedia Modeling, pp. 39–46 (1998)
7. Malik, O.A., Ismail, N., Hussein, B.R., Yahya, U.: Automated real-time identification of medicinal plants species in the natural environment using deep learning models—a case study from Borneo Region. Plants **11**(15), 1952 (2022)
8. Valdez, D.B., Aliac, C.J.G., Feliscuzo, L.S.: Medicinal plant classification using convolutional neural network and transfer learning. In: 2022 IEEE International Conference on Artificial Intelligence in Engineering and Technology (IICAIET), pp. 1–6. IEEE (2022)
9. Abdollahi J.: Identification of medicinal plants in ardabil using deep learning: identification of medicinal plants using deep learning. In: 27th International Computer Conference, Computer Society of Iran (CSICC), pp. 1–6. IEEE (2022)
10. Zin, I.A., Ibrahim, Z., Isa, D., Aliman, S., Sabri, N., Mangshor, N.N.: Herbal plant recognition using deep convolutional neural network. Bull Electr. Eng. Inform. **9**(5), 2198–2205 (2022)
11. Saikia, A.P., Hmangaihzuala, P.V.L., Datta, S., Gope, S., Deb, S., Singh, K.R.: Medicinal plant species classification using neural network classifier. In: 2021 6th International Conference on Communication and Electronics Systems (ICCES), pp.1805–1811. IEEE (2021)
12. Manoharan, J.S.: Flawless: detection of herbal plant leaf by machine learning classifier through two stage authentication procedure. J Artif Intell Capsule Netw. **3**(2), 125–139 (2021)
13. Haque, M.A., Marwaha, S., Deb, C.K., et al.: Deep learning-based approach for identification of diseases of maize crop. Sci. Rep. **12**, 6334 (2022)
14. Hu, H., et al.: The identification of fritillaria species using hyperspectral imaging with enhanced one-dimensional convolutional neural networks via attention mechanism. Foods (2023)
15. Tariku, G., Ghiglieno, I., Gilioli, G., Gentilin, F., Armiraglio, S., Serina, I.: Automated identification and classification of plant species in heterogeneous plant areas using unmanned aerial vehicle-collected rgb images and transfer learning. Drones **7**, 599 (2023)
16. Lee, C.P., Lim, K.M., Song, Y.X., Alqahtani, A.: Plant-CNN-ViT: plant classification with ensemble of convolutional neural networks and vision transformer. Plants **12**, 2642 (2023)

17. https:// www. kaggle. com/ datas ets/ vishn uoum/ medic inalplantdataset augmented? select= data
18. Sediqi, K.M., Lee, H.J.: A novel upsampling and context convolution for image semantic segmentation. Sensors. **21**, 2170 (2021)
19. htttps://learnopencv.com/deeplabv3-ultimate-guide/
20. Cherian, I., Agnihotri, A., Katkoori, A.K., Prasad, V.: Machine learning for early detection of alzheimer's disease from brain MRI. Int. J. Intell. Syst. Appl. Eng. **11**(7s), 36–43 (2023)
21. Kumar, K.A., Boda, R.: A computer-aided brain tumor diagnosis by adaptive fuzzy active contour fusion model and deep fuzzy classifier. Multimed. Tools Appl. **81**, 25405–25441 (2022)
22. Dudi, B., Rajesh, V.: A computer aided plant leaf classification based on optimal feature selection and enhanced recurrent neural network. J. Exp. Theor. Artif. Intell. **35**(7), 1001–1035(2022)

Android App Permission Detector Based on Machine Learning Models

Jaikishan Mohanty and Divyashikha Sethia(✉)

Department of Software Engineering, Delhi Technological University, Delhi, India
divyashikha@dtu.ac.in

Abstract. Android has dominated a significant portion of the cellphone market as one of the most developed intelligent operating systems for mobile devices. However, the issue of Android malware detection remains critical due to the security mechanism and the lack of stringent validation during the publishing of Android apps, leading to potential breaches of user privacy through unwanted permissions. As mobile app security and privacy remain critical, integrating the machine learning model into Android applications offers a practical solution. This paper proposes an Android application using machine learning models. The application provides a user-friendly interface to assess the permissions requested by various apps and recommends the appropriate actions based on the permissions' safety ratings. The model focuses on the safety of app permissions based on their usage frequency in specific app categories available on the Google Play Store. Subsequently, the permissions are evaluated and rated as either safe or unsafe. This paper contributes to a safer mobile app ecosystem by providing users with a comprehensive tool for making informed app permissions decisions.

Keywords: Android · app permission · classification · machine learning · android security · malware detection

1 Introduction

With its ever-expanding user base, the Android ecosystem has grown remarkably in recent years. According to IDC, in 2022, Android stands as the dominant player in the global mobile operating system arena. It holds a market share of around 84.2% throughout mobile operating systems worldwide [1]. With more than 3 billion active users globally and growing, this operating system's popularity is on the rise, largely thanks to the accessibility of low-cost smartphones and the expansion of mobile internet access [2]. The Google Play Store releases apps often, which raises the possibility of abnormal behaviour like data access without the user's consent. Virus threats are also rising along with an increase in the number of apps available in the Google Play Store. Keeping track of all the third-party apps offered in the Google Play Store takes time and effort. The Google Play Store implements security safeguards and often removes apps, but malware publishers continually look for new vulnerabilities and republish the apps.

However, the rise of Android malware has become a significant concern for users and developers. The security mechanism of Android and the lack of stringent validation during app publishing have contributed to the proliferation of malware on the platform [3]. Malware poses a significant threat to user privacy and security, as it can gain unauthorized access to personal information and sensitive data.

Android presents several privacy threats [4], such as:

- *Malware*: Without the user's knowledge, malicious software or malware is installed on an Android device, gathering private data, including login passwords and financial information.
- *Location tracking*: A few applications can access and track a user's location without that user's knowledge or consent.
- *Data Leakage*: Android apps can access various private data, including contacts, location, and text messages, and they may do so without the user's knowledge or approval.
- *Phishing*: Phishing attacks target Android users, tricking them into providing sensitive information, including login credentials, to a bogus website or app.

Applications must obtain user permission before accessing specific data and capabilities under Android's permission system. However, it is ultimately up to the users to allow or deny these permissions, which is problematic. More than 70% smartphone apps often request data collection permissions that are only sometimes essential for their core functions. Unfortunately, many users hastily grant all permissions during installation, while even those who pay attention may need help comprehending the implications fully. The lack of practical guidance for granting permissions leaves users needing assistance navigating the minefield of app permissions. The consequences are dire, as third-party apps can covertly track user locations, access sensitive information like contact lists and SMS logs, and compromise personal data privacy [5]. With Android 6.0, Android introduced runtime permissions to address these issues and give users more control over app permissions. There are four types of permissions: install-time permissions, runtime permissions, normal permissions, and special permissions. These permissions define the scope of data access for an app and the actions the app can perform if granted permission.

1. *Install-time permission:* It provides restricted access to classified data or actions that have a minimal impact on the system. The user receives a notification when they view the app's information page, notifying them of the install-time permissions. Examples of these permissions include ACCESS_NETWORK_STATE, BLUETOOTH, and INTERNET.
2. *Runtime permission:* It grants the app additional access to classified data or actions that can significantly impact the system and other apps. Therefore, the app must request runtime permissions before accessing classified data or performing restricted actions. Examples of these permission groups include Calendar, Call log, Camera, Contacts, Location, Microphone, Sensor, SMS, Storage, and Telephone.
3. *Special permission:* These permissions are established when the platform and Original Equipment Manufacturers(OEMs) wish to restrict access to decisive actions such as drawing over other apps. The Special app access page in system settings holds a collection of user-toggle-able operations, with many implemented as special permissions. These permissions include scheduling exact alarms, drawing over other apps, and Accessing all storage data.

Runtime permissions are also known as user authorization permissions, primarily encompassing permission groups. However, none of the Android versions provide any service to warn the user about the safety of the application permissions. The app permissions management system has changed since the recent Android 12 update [6]. The update's notification service covers only a small number of specific permissions. Other permissions may also pose a threat to privacy. New malware techniques developed for new apps mainly threaten users' privacy. Therefore, there is a need for new Android services that can warn users when something is safe or unsafe and give them a secure environment in which to use applications.

Android 13 [7] brings several vital updates. Users can now stop foreground services from the notification drawer, improving performance and battery life. A new runtime permission, POST_NOTIFICATIONS, enhances user control over notifications. Apps targeting Android 13 must declare the AD_ID permission for Google Play services. Privacy and security need separate media access permissions instead of READ_EXTERNAL_STORAGE. Install-time permission USE_EXACT_ALARM benefits apps like calendars and alarms, ensuring precise timing. These changes reflect Android's Focus on performance, privacy, and security.

To address this problem, APEC, also known as App Permission Classification with Efficient Clustering [8], is a model that aims to enhance the security of Android applications. It utilizes a three-tier structure consisting of permission categorization clustering through DBSCAN and permission classification using Decision Tree and Random Forest algorithms. APEC achieves accuracy rates, with Decision Tree reaching 93.8% accuracy and Random Forest achieving 95.8% accuracy. Moreover, it effectively addresses privacy concerns by predicting whether permissions are safe or unsafe based on the categories of the apps. This model offers recommendations for users and developers, improving Android app security in line with the updates introduced in Android 12 [6] for permission management.

Contribution: The main contributions of the proposed work include:

- *Proposal of an Android App Permission Detector*: This paper proposes a novel Android App Permission Detector for Permission classification based on a static model APEC [8]. The application checks whether the application permissions are safe or unsafe based on their usage frequency in the app category on the Google Play Store. The application aligns with Android 12 [6] and Android 13 [7] updates and outperforms similar apps. The application acts as a user recommendation system. It suggests to developers the minimal permissions required for the app to work perfectly and for users to give information about the unsafe permission for every installed application.
- *Testing of Proposed Android Permission Detector*: The Proposed Android Permission Detector application undergoes continuous rigorous testing on the various real-time apps from the Google Play Store for the different cate- gories across different Android platforms, ensuring it aligns with the evolving expectations of users and developers.

Roadmap: The rest of the paper is structured clearly and concisely. Section 2 provides related work, highlighting the previous research efforts and comparing them with the existing applications and Android versions in this domain. Section 3 defines the Overview of the APEC Model [8]. Section 4 outlines the Implementation of the APEC model in an Android application. Section 5 presents the Results and Discussion. The paper concludes with Section VI, Conclusion and Future Work.

2 Related Work

In 2022, Manzil et al. [9] proposed a detection framework based on permission features using machine learning techniques and Recursive Feature Elimination (RFE) technology. The system analyzes 100 CoVID-themed fake apps from the Google Play Store and Github repository and extracts permission features from the AndroidManifest.xml file. The system consists of 4 components: dataset collection, static analysis, feature selection, and classification. The study shows better accuracy with the Decision tree and random forest classifiers. Most malware identified are part of the Cerberus, SpyNote, Metasploit payloads, and SMS Stealer family. This framework may not detect new or unknown variants of COVID-themed Android malware that researchers have yet to include in the dataset.

SAMADroid, a novel three-level hybrid malware detection model for the Android operating system, was proposed by Arshad et al. (2018) [10]. The authors conduct an analysis and categorize several Android malware detection methods. They developed a revolutionary three-layer hybrid malware detection methodology for the Android operating system that combines the advantages of three layers: static and dynamic analysis, local and remote hosts, and machine learning intelligence. According to trial data, SAMADroid ensures efficiency in terms of power and storage consumption to reach a high level of accuracy.

Sun et al. (2016) [11] proposed a method called Significant Permission Identification for Android Malware Detection (SIGPID). This method aims to tackle the malware issue on Android devices by identifying the required crucial permissions. Their study showed that SIGPID outperformed techniques in terms of effectiveness. It achieved an accuracy rate of 93.62% for detecting apps. It has demonstrated a classification accuracy of 91.4% when tested with unknown malware apps.

Table 1 contrasts the related work's main contributions and limitations. The earlier studies demonstrate that the dataset needed to be revised and unjustifiable given the variety of apps available in the Google Play Store. Additionally, the prior efforts' main emphasis was on malware identification rather than the privacy issues brought on by the programs' additional permission requests. Therefore, a new approach is needed to solve the privacy and data abuse issues brought on by the apps' other permission requests.

2.1 Comparison with the Existing Application

Mobile app permissions are crucial for user security and privacy. Applications have evolved to manage permissions, allowing users to turn them on or off. Malware exploiting app permissions is a growing threat to user privacy.

This paper delves into a comparative analysis of the "Permission Detector" application and several similar applications in the domain of permission management. It provides

Table 1. Related Work On Model

Research Work	Type	Analysis Tech. Used	Contribution	Limitation
Manzil et al. (2022) [9]	Malware Detection	Static Analysis	It identifies COVID-themed Android malware and improves accuracy using Recursive Feature Elimination and machine learning.	The dataset was custom-created using feature elimination techniques and may not detect new or unknown variants of COVID-themed Android malware not in the dataset.
Sun et al., (2017) [11]	Malware Detection	Static Analysis	Multilevel Data Pruning	Out of 122 possible permissions, the model only takes 23 into account. False classification may be the outcome of the reduction.
Arshad et al. (2018) [10]	Malware Detection	Static and Dynamic Analysis	SAMADroid have shown a slight performance overhead on Android smartphones, outperforming the work Drebin dataset in static analysis.	The local host does not detect malicious activity, resulting in the failure of network links or congestion at the channel.

a comprehensive view of the functionality, limitations, and malware detection capabilities of Permission Detector compared to other permission management applications. This comparative analysis aims to highlight the distinctive strengths and contributions of Permission Detector in enhancing mobile app security and user privacy.

The Permission Manager [12] primarily focuses on enabling or disabling specific permissions for user-installed and system applications. However, it operates with limited permissions, including Calendar, Call log, Camera, Contacts, Location, Microphone, Sensor, SMS, Storage, and Telephone. Notably, it cannot modify sensor permissions starting from Android 12. Despite its functionality, Permission Manager does not address malware detection through permission classification.

Similarly, Permission Pilot [13] presents a dashboard listing all installed applications and their permissions from the Manifest.xml file. Users have control over viewing and disabling granted and ungranted permissions, but like Permission Manager, it lacks malware detection capabilities based on permission classification.

On the other hand, App Permission Manager [14] provides information about installed, system, and high-risk apps, specifying granted permissions from various categories. While identifying high-risk apps and considering granted permissions without evaluating their safety. Like the other tools, App Permission Manager does not feature malware detection based on permission classification.

Table 2 describes that other applications provide valuable functionality for managing app permissions. The proposed Android application Permission Detector stands out with its unique capability for malware detection through permission classification. It empowers users to not only control permissions but also assess the safety of these permissions, contributing to a more secure and privacyconscious mobile app ecosystem.

2.2 Comparison with the Android Versions

Android released a new update with significant changes in the app permission management. Here is a comparison of updates in the app permission management with the proposed application as shown in Table 3 .

Table 2. Comparison of Android

Aspects	Number of permissions	Purpose	Safe/Unsafe support
Permission Manager [12]	Only user authorizes permissions	Enable and Disable permission.	No
Permission Pilot [13]	All permissions of AndroidManifest.xml file.	Display Granted and ungranted permissions, and the user can disable it from the setting.	No
App Permission Manager [14]	Only user authorizes permissions	Presents high-risk apps based on four granted permissions: Contact, SMS, Telephone, and Storage.	No
Proposed Android Permission Detector	All Runtime and install time granted permissions.	Detect safe or unsafe permission based on their usage frequency in the app category.	Yes

3 Overview of APEC: App Permission Classification with Efficient Clustering [8]

The APEC model [8] is an approach for classifying app permissions into safe and unsafe categories using a three-tier architecture. It uses a 2 million dataset.

comprising the name of the app and all app permissions requested by the applications. It efficiently categorizes app permissions based on the frequency of their occurrence within different app categories. The key components of the APEC model, as shown in Fig. 1, comprise of Group category, Clustering and Approval and Classifier.

1. *Group Category*: The APEC model categorizes apps based on their respective categories in Layer 1, as shown in Fig. 1. This categorization stems from apps within the same category generally sharing similar requirements and permissions, as explained in [15]. To group apps by category, the APEC model calculates the sum of each unique permission in every category on the basis requested by the apps. It then creates a frequency map of permissions based on the category containing the sum of each permission.

Table 3. Comparison with Android Versions

Aspects of Comparison	Selective Permission Control	Scope of Permission Types	Privacy Concerns	Comprehensive Coverage
Android 12 [6]	Selective notification for microphone and camera access only.	Focus on platform-provided permissions, excluding runtime and special permissions.	Permission groups might only address some privacy concerns and variations among apps.	Limited scope, concentrating on a few specific permissions.
Android 13 [7]	User can stop Foreground service from the Notification drawer.	Some of the runtime permissions covered, e.g., POST_NOTIFI-CATION	Separate permission for EXTERNAL_-STORAGE	Calendar and Alarm introduce some new permissions, i.e., USE_EXACT_-ALARM at install time.
Proposed Permission Detector	Notifications cover all user permissions.	Consider a wide array of permissions, including runtime and special ones.	Permission categories ensure better addressing of various privacy concerns.	It covers a wide range of permissions to provide comprehensive coverage.

2. *Clustering and Approval*: The second tier of the APEC model uses DBSCAN clustering to group permissions based on their frequency in Layer 2, as shown in 1. DBSCAN is a density-based spatial clustering of applications with a noise algorithm well-suited for clustering data with outliers. The APEC model uses DBSCAN clustering to group permissions into core and outlier clusters. The core cluster contains the permissions most frequently requested by apps in a category, while the outlier cluster contains the less frequently requested permissions.

To determine which permissions belong to the core cluster, the APEC model calculates the k-distance for each permission. The k-distance is the minimum distance between permission and its kth nearest neighbour. The APEC model then plots the graph of k-distance values and identifies the point where the line curves. Figure 2 shows the k-distance graph of the action category. It shows the line approximately curves at a value of $\varepsilon = 20$. Hence, in DBSCAN clustering, the value will be in the neighbourhood of 20 to give an optimal clustering of the app permissions of the action category. This point is considered the optimal epsilon value, and the cluster uses it for evaluation.

The APEC model also assigns an approval rating to each permission. The approval rating is represented by a '1' for the cluster's core, showing that the permission is safe,

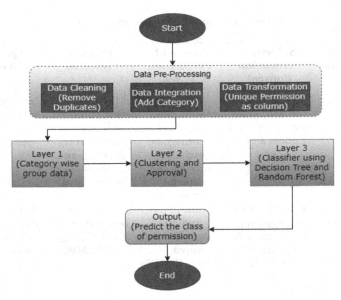

Fig. 1. Flow diagram of Three-Tier Architecture for APEC [8]

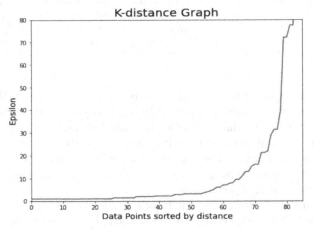

Fig. 2. K-distance graph of Action category [8]

and an unsafe '0' for the cluster's outlier, showing each category's rare permissions requests.

3. *Classifier*: The third tier of the APEC model [8] uses a decision tree or random forest classifier to evaluate the permissions of new apps according to their category and the approval rating of the permissions within that category. Tier 2 clustering generates ground truth data and trains the decision tree or random forest classifier. To evaluate the permissions of a developed application, the APEC model follows a step-by-step

process. Firstly, it idenifies the category that the app belongs to. Then, it assigns an approval rating for each permission associated with that category. Lastly, using either a decision tree or random forest classifier, it predicts the safety level of each permission by considering its approval rating.

4 Implementation of APEC in the Android Permission Detector

An Android application that implements the APEC model [8] allows users and developers to assess the security of app permissions. The application provides a user-friendly interface to assess the permissions requested by various apps and recommends the appropriate actions based on the permissions classification.

Integrating the APEC model [8] into the Android application is seamless, providing users with a straightforward and user-friendly experience. This integration allows users to access the APEC model's functionality without complexity. The integration of the APEC model [8] into the Android application, as shown in Fig. 3, involves several key steps:

1. *Dataset*: The work enhances the dataset used in the APEC [8] Model by extracting the model predicted output and integrating it into the Android application known as Permission detector.
2. *Initial Installation and Data Collection*: Upon the initial installation of the application, it queries all application packages and Loaded APEC model results into an application.
3. *Pre-processing Dataset*: The collected dataset of query packages undergoes pre-processing to eliminate all system applications in the list. This step ensures that the analysis focuses on user-installed applications only.
4. *Exporting Permissions*: Users can export all user-installed apps and their granted permissions into an Excel file. This functionality supports future permission classification, data analysis, and app security work.
5. *Fetching Granted Permissions*: When the user clicks on any application, it fetches all the granted permissions by checking if PackageInfo.requestedPermission = = PackageManager.PERMISSION_GRANTED, then add in a list, forming the basis for further analysis.
6. *Comparison with APEC Model*: Compared the granted permissions with the APEC model's [8] results. This comparison categorizes permissions as safe or unsafe, providing users with clear insights into their security.
7. *Enable or Disable Permission*: Users can interact with the application through a user-friendly interface. They can turn specific permissions on or off directly from the application, giving them greater control over their device's security.

5 Results and Discussion

The results of this study demonstrate the successful integration of the App Permission Classification with Efficient Clustering (APEC) Model [8] into an Android application for permission classification, allowing users to distinguish between safe and unsafe app permissions. Notably, the model achieved an impressive accuracy rate of 93.8% using Decision Tree and 95.8% using Random.

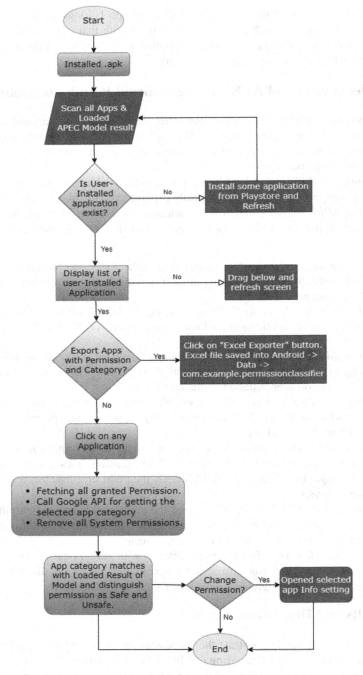

Fig. 3. Flow diagram of Android Permission Detector

Forest classifiers. The DBSCAN clustering algorithm, an unsupervised learning methodology, effectively rated apps within various categories. The classification model, trained on 100% of data from a dataset comprising two million apps and tested with a real-time Google Play Store application, empowers users to manage permissions, safeguarding their privacy.

Table 4. Application Testing with Proposed Application

Device / Android Version	Category	App Name	Unsafe Permissions	Unsafe Permissions that Users Can Enable/Disable
LG / 12	Books and Reference	Freed Audiobooks	WAKE_LOCK FOREGROUND_SERVICE WRITE_EXTERNAL_STORAGE ACCESS_COARSE_LOCATION ACCESS_FINE_LOCATION	WRITE_EXTERNAL_STORAGE ACCESS_COARSE_LOCATION ACCESS_FINE_LOCATION
Samsung / 10	Beauty	TST-Malaysia	CHANGE_NETWORK_STATE READ_CONTACTS WRITE_EXTERNAL_STORAGE CALL_PHONE WAKE_LOCK GET_TASKS	READ_CONTACTS WRITE_EXTERNAL_STORAGE CALL_PHONE
POCO / 13	Board	Fun101 Okey	WRITE_EXTERNAL_STORAGE RECORD_AUDIO MODIFY_AUDIO_SETTINGS WAKE_LOCK KILL_BACKGROUND_PROCESS FOREGROUND_SERVICE	WRITE_EXTERNAL_STORAGE RECORD_AUDIO

Fig. 4. The interface of Permission Detector. In Fig. (a), Audiobooks show granted unsafe permission, of which three are user-authorized. In Fig. (b), After turning off User-authorized permission from setting and refresh the app.

Testing on Google Play Store applications across all 48 different categories [16] revealed the accuracy for determining safe and unsafe permissions, enabling users to make informed decisions regarding permission management. Furthermore, rigorous testing on various models of Android phones, including the LG G8X, Motorola, Realme, Redmi, OnePlus 6, POCO, OnePlus, and Samsung, demonstrated the versatility and robustness of the system. These results underscore the system's effectiveness across various Android devices, enhancing its applicability in real-world scenarios. For example, as depicted in Fig. 4, the Freed Audiobooks application determines the granted unsafe permissions for which the user authorizes only a few. After turning off user-authorized permissions, the Permission Detector app removes those unsafe permissions. This paper displays testing results in Table 4, and you can locate the complete results in [17]. These results underscore the potential of the APEC Model [8] in bolstering mobile app security and user privacy while fostering transparency between users and developers.

6 Conclusion and Future Work

Android is the most popular operating system in the world. App permissions are essential to strengthening the Android ecosystem's security. Android is constantly improving the robustness of app permissions. However, the vast user base has made the Google Play Store vulnerable to several fraudulent applications, compromising its security. The Android ecosystem underwent a substantial transformation in permission management with the advent of the Android 12 and 13 updates. Despite these improvements, no Android version currently provides a mechanism to proactively alert users regarding the safety or potential risks associated with application permissions. Additionally, while existing applications offer valuable functionality for managing app permissions, they do not assess the safety implications of these permissions. This paper introduces a novel Android App Permission Detector, grounded in a static model known as APEC [8]. Incorporating the APEC model into an Android application marks a significant stride towards augmenting the security of mobile app permissions. By offering users a user-friendly tool for evaluating and managing app permissions, this approach empowers individuals to assert control over their digital privacy and security. The proposed application has undergone comprehensive testing on various applications spanning all categories available on the Play Store across different Android platforms.

In the future, the Android App Permission Detector can improve permission classification to address emerging app categories, and extending the application to provide post-installation security alerts from the Play Store will empower users to make informed decisions, ensuring a more secure and privacy-aware mobile app ecosystem.

References

1. Nabila Popal, A.S.: IDC smartphone market sharemarket share. https://www.idc.com/promo/smartphone-market-share. Accessed 15 Sep 2023
2. IANS, Android now powers over 3 billion devices worldwide. https://bit.ly/android_devices_sundar_pichai . Accessed 07 Oct 2023

3. X. Su, et al.:An informative and comprehensive behavioral characteristics analysis methodology of android application for data security in brain-machine interfacing. https://scite.ai/reports/10.1155/2020/3658795. Accessed 10 Oct.2023
4. Yang, et al.: Pradroid: Privacy risk assessment for android applications. In: 2021 IEEE 5th International Conference on Cryptography, Security and Privacy (CSP), pp. 90–95 (2021)
5. Olmstead, K., Atkinson, M.: An analysis of android app permissions. pew research center: Internet, science & tech. https://bit.ly/pewresearch_android. Accessed 12 Oct 2023
6. Google, Android 12 Features Overview. https://bit.ly/3QK9q3t. Accessed 1 Oct 2023
7. Google, Android 13 Features Overview. https://bit.ly/40pNqOz. Accessed 18 Oct 2023
8. Rawal, P.S, Sethia, D.:APEC: App Permission classification with Efficient Clustering. In: 2023, International Conference On Computational Intelligence For Information, Security And Communication Applications (CIISCA) (Accepted in conference) (2023)
9. Manzil, et al.: Covid-themed android malware analysis and detection framework based on permissions. In: 2022 International Conference for Advancement in Technology (ICONAT), pp. 1–5 (2022)
10. Arshad, S., et al.: Samadroid: a novel 3-level hybrid malware detection model for android operating system. IEEE Access **6**, 4321–4339 (2018)
11. Sun, L., et al.: Sigpid: significant permission identification for android malware detection. In: 2016 11th International Conference on Malicious and Unwanted Software (MALWARE), pp. 1–8 (2016)
12. Permission Manager created by NorthRiver (2019). https://play.google.com/store/apps/details?id=com.agooday.permission
13. Permission Pilot created by darken (2022). https://play.google.com/store/apps/details?id=eu.darken.myperm details?id=eu.darken.myperm
14. App Permission Manager created by Micro Inc (2021). https://play.google.com/store/apps/details?id=com.assistant.android.permission.manager
15. Felt,A., et al.: Android permissions: user attention, comprehension, and behaviour. In: SOUPS 2012 Proceedings of the 8th Symposium on Usable Privacy and Security (2012)
16. Google, Android category and tags for your app or game. https://bit.ly/android_category . Accessed 05 Oct 2023
17. Android Permission Detector Testing Results. https://bit.ly/testing_result

Histopathological Analysis Advancements in Deep Learning for the Diagnosis of Lung and Colon Cancer with Explanatory Power via Visual Saliency

Seema Kashyap[1], Arvind Kumar Shukla[1](✉), and Iram Naim[2]

[1] School of Computer Science and Engineering, IFTM University, Uttar Pradesh, Moradabad, India
`arvindshukla@iftmuniversity.ac.in`
[2] Faculty of Engineering and Technology, MJP Rohilkhand University, Bareilly, India
`iram.naim@mjpru.ac.in`

Abstract. One of the most dangerous diseases in modern times is generally thought to be lung cancer. Colon cancer will be the second biggest killer of Americans in 2023 when it comes to cancer. The most precise diagnosis technique involves the utilisation of histopathology pictures obtained from biopsies. Histopathological image analysis is the process of carefully looking through tissue samples from people with lung and colon cancer to find cancerous cells and tumours. This research is very important for figuring out the stage of the tumour, which makes diagnosis easier, predicts the prognosis, and helps plan treatment. Lung and colon cancer are significant public health issues that require continuous efforts in prevention, early identification, and effective therapeutic options. This approach employs deep learning, namely the EfficientNet architecture. The study employs several image resolutions and applies transfer learning and parameter tweaking techniques to improve performance. Utilising the LC25000 dataset, the proposed approach attains a notable degree of precision, with EfficientNetB0 emerging as the most accurate model, achieving a flawless score. The study emphasizes the ability of deep learning to automate the classification of cancer-related histopathology images, hence improving the efficiency and accuracy of diagnostic operations. The study addresses the inherent black box nature of deep learning (DL) models by emphasizing the importance of explainability in understanding their predictions. To visualize and interpret the decision-making process of EfficientNetB0 predictions, the study employs lime, a method known for generating visual saliency maps. This allows to identify and highlight the most activated areas in input images, providing valuable insights into the features influencing the DL model's class predictions.

Keywords: histopathology pictures · colon and lung cancer · EfficientNetB0 · Explainable AI

1 Introduction

Cancer is a collection of diseases characterized by spontaneous genetic abnormalities in cells, resulting in the development of aberrant cells that proliferate without control and have the ability to metastasize to various organs. Cancer prediction traditionally depends on clinical information, but in recent times, many data sources, including biopsies and imaging scans, have been employed to enhance our comprehension of the illness. Lung and colon cancers are identified as significant factors in global cancer mortality. The estimated number of people diagnosed with lung cancer in the US in 2023 is 238,340, with 117,550 men and 120,790 women affected. The American Cancer Society reports that in 2023, cancer of the intestines will be the second most common cancer killer in the US. The yearly estimate is 153,020 cases of CRC diagnosis and 52,550 deaths. The study examines the utilization of the EfficientNetB0 algorithm for the classification of colon and lung cancer. The study highlights the crucial need of promptly identifying and precisely categorizing cancer for effective therapy. To ensure uniformity, the study employs publicly accessible datasets containing pre-processed photos. EfficientNetB0 exhibits superior accuracy, precision, recall, and F1 scores on all cancer datasets, surpassing other cutting-edge algorithms in the existing research. EfficientNet is a neural network architecture that aims to provide excellent performance in image classification tasks while keeping computing complexity to a minimum. It achieves this by systematically increasing the depth, width, and resolution of a baseline architecture using a compound scaling algorithm. The learning rate is an essential hyperparameter in deep learning, which determines the magnitude of adjustments made to a model during training. Choosing an optimal learning rate is a complex task, as excessively small or large numbers can result in training difficulties. Although there is continuous study and contributions towards analysing adaptive learning rates, the task of determining an appropriate rate still remains unresolved. This is because it depends on the characteristics of the data and the specific problem being addressed. Lung and intestinal cancer histology slide images are classified using efficientnetB0 as the CNN model and CLAHE. The primary focus is on image preprocessing, particularly utilising CLAHE, to attain colour equilibrium and preserve overall intricacies. Preprocessing is essential for achieving optimal feature learning in the classification of histopathology images.

2 Literature Review

Multiple researchers utilize diverse neural network and deep learning methods for the purpose of categorizing and identifying patterns, with a specific focus on artificial intelligence systems that rely on Convolutional Neural Networks (CNN). Recent years have seen CNN use in several fields. Using the LC25000 histopathology imaging dataset, which was published in 2020 and is specific to colon and lung cancer, the evaluation is performed. This section examines numerous researchers who used this dataset to construct deep learning applications.

Hamed et al. [2] suggested the CNN-LightGBM method for sorting histopathological lung cancer datasets and talked about how well deep learning techniques work in general for medical imaging and cancer diagnosis.LC25000 dataset of the Lungs and Colon cancer is used in this research and achieved 99.6% accuracy and sensitivity. Martínez et al. [3] suggested the LR strategy with the VGG19 model, which can be used to learn and test the strategy on various cancer datasets. Using the triangular-cyclic learning rate, they were able to make a system that was 96.4% correct on test images for nine different types of tissue. Sudhakar Tummala et al. [4] created a process that used EffcientNetV2 models that had already been trained to accurately sort five groups into groups with a 90% success rate. These results show that deep learning has a lot of promise for automatically classifying different types of cancer. The dataset that was used in the study was LC25000. The study employs GradCAM, a method known for generating visual saliency maps. This allows to identify and highlight the most activated areas in input images, providing valuable insights into the features influencing the DL model's class predictions. Md Sabbir Ahmed et al. [5] created an interpretable lung cancer detection system using four ML and two XAI models. The study found Logistic Regression and Random Forest to predict lung cancer best. These models improved interpretability and classification acceptance when paired with SHAP and LIME. The study prioritised lung cancer diagnosis precision and explainability. Sugondo Hadiyoso et al. obtained 98.96% accuracy in the automatic classification of lung and colon cancer, demonstrating the effectiveness of CLAHE and the promise of deep learning and image processing in

Table 1. Convolutional models used in the study of the Lung Cancer Dataset

author	model	dataset	accuracy
Hamed et al. [2]	CNN-LightGBM method	LC25000	99.6%
Martínez et al. [3]	LR strategy with the VGG19 model	LC25000	96.4
Sudhakar Tummala et al. [4]	EffcientNetV2	LC25000	90%
Md Sabbir Ahmed et al. [5]	Logistic Regression and Random Forest	LC25000	97%
Sugondo Hadiyoso et al	CNN + CLAHE	LC25000	98.96%
Qasthari, B. L et al. [7]	vgg19 and AI technologies	LC25000	99.96% with 1.5% loss rate
Masud et al. [8]	CNN + 2D-DFT,2D-DWT image transforms	LC25000	96.6
Ahmed S. Sakr et al. [9]	Lightweight CNN model	LC25000	99.5 for colon cancer
Nora Yehia et al. [10]	efficientnetB7	LC25000	99.5
Proposed model	CNN model using pretrained efficientnetB0 + clahe	LC25000	100

cancer picture classification. Author achieved a 98.96% accuracy in automated lung and colon cancer categorization. Qasthari, B. L et al. [7] used vgg19 and AI technologies to create a very accurate model for classifying histological images of lung and colon cancer. Masud et al. [8] developed a framework for classifying histological cancer images using Convolutional Neural Networks (CNNs). In order to classify pictures, four sets of features were combined by concatenating the outcomes of applying 2D-DFT and 2D-DWT image transformations. The proposed technique for identifying lung and colon cancer attained a peak classification accuracy of 96.33% and an F-measure score of 96.38%. Ahmed S. Sakr et al. [9] introduced a lightweight deep learning approach that utilised Convolutional Neural Networks (CNNs) to detect colon cancer. This method demonstrated an exceptional accuracy of 99.50%. The suggested deep learning model's accuracy, precision, recall, and F1-score for detecting colon cancer were evaluated using LC25000 lung and colon histopathology images.Nora Yehia et al. have created an AI classification system using efficientnetB7 that can accurately spot cancerous tissues 99.5% of the time, showing big improvements in performance across both training and testing datasets. On the LC25000 dataset, this system had 99.6% precision, 99.4% recall, and a 99.6% F1-score (Table 1).

3 Methodology

3.1 Dataset

The LC25000 Lung and Colonic histological image collection, comprising 15,000 histopathological images of lung cancer and 10,000 histopathological images of colon tissue, was employed to evaluate the efficacy of our research. Images are categorized into five distinct classes. The dimensions of all images are 768 × 768 pixels, and they are saved in the JPEG format. The training dataset consisted of 20,000 pictures, while the validation and testing datasets were 2,500 photos each (Figs. 1, 2).

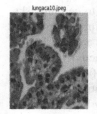   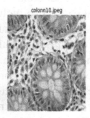 

Fig. 1. image dataset example.

image Label	type of image	number of images
Colon_aca	colon Adenocarcinoma	5000
colon_n	colon Benign tissue	5000
Lung_aca	lung Adenocarcinoma	5000
Lung_n	lung Benign tissue	5000
Lung_scc	lung Squamous-cell carcinoma	5000

Fig. 2. image dataset distribution in each class.

The Proposed Approached

- Specifies the dimensions of the image, the size of the batch, and various other characteristics.
- Calculates optimal test batch size and steps.
- Improves image contrast by utilising the LAB colour scheme, histogram equalisation, and CLAHE.
- Normalises pixel values to a range of -1 to 1.
- Generates data sets for training, testing, and validation by utilising Keras' ImageDataGenerator.
- Use pretrained efficientnetB0 model to classify the image.
- Use lime and Shap for image visualization

The Structural Design of the Proposed Methods

The study utilises a model architecture that employs a pre-trained EfficientNetB0 base for extracting features. It also includes additional layers for adapting to certain tasks. The architecture incorporates global max pooling to reduce the spatial dimensions, batch normalisation for normalisation purposes, a dense layer with 256 units and ReLU activation function, and regularisation techniques such as L2 kernel, L1 activity, and bias regularisation. The model incorporates a dropout layer with a rate of 0.45 to improve regularisation. The selected activation function for multi-class classification is SoftMax. Adamax optimizer with 0.001 learning rate and category cross-entropy loss function is used to optimise the model. The assessment is conducted using the accuracy metric. This architectural design integrates pre-trained features with task-specific adaptation and regularisation, resulting in efficient multi-class classification on the provided dataset (Fig. 3).

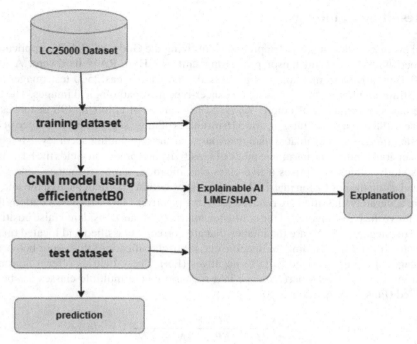

Fig. 3. Model architecture of proposed methodology.

A thorough evaluation of the model's ability to describe the test data, showing how well it can generalize and put things into different groups. The printed report gives a full picture of the model's accuracy, recall, and F1-score in different groups. The confusion matrix using a model's forecasts on a test dataset and shows the plot that comes up as a result. The visualization helps you understand how well the model sorts things into different groups by showing you which guesses were right and which were wrong. The model use LIME to elucidate the predictions of an image categorization. This tool produces and displays a saliency map for the highest predicted label, showing the areas in the original image that have the greatest impact on the model's conclusion. The model utilizes the SHAP (SHapley Additive exPlanations) library to calculate and display SHAP values for a picture. The shap values quantify the influence of individual pixels on the output of the model. The SHAP (Shapley Additive Explanations) method is utilized to generate an image plot, which provides valuable insights into the influence of individual pixels on the decision-making process of the model. This aids in the interpretation of the model's behavior.

4 Result and Findings

Model evaluating the suggested approach by utilizing the Google Colab Pro application equipped with a powerful tensor processing unit and High-RAM hardware. A total of 25000 photos were meticulously processed. The dataset has 2500 test images and 20,000 training photos encompassing all cancer-type histopathological images. The test image has 479 instances of colon _ACA, 519 instances of colon_n, 498 instances of lung_aca, 488 instances of lung_n, and 516 instances of lung_scc. The performance of the suggested models was evaluated using various measures, including accuracy, F1-score, and precision. Finding the harmonic mean of sensitivity and precision yields the F1-score. In this study, which comprises a five-class classification, the performance scores are obtained using the confusion matrix. More precisely, within a particular class, the images that are accurately classified are referred to as true positives (TP). Misclassifications that occur above the half-diagonal of the confusion matrix (CM) are considered false positives (FP). True negatives (TN) are the images that are correctly classified and located on the diagonal of the CM, excluding the specific class. Misclassifications that occur below the half-diagonal are categorized as false negatives (FN). These metrics offer a thorough evaluation of the model's performance in a scenario where multiple classes are being classified (Figs. 4, 5, 6, Table 2, 3).

$$Accuracy = \sum_c \frac{TP_c + TN_c}{TP_c + FP_c + TN_c + FN_c}, c \in classes \tag{1}$$

$$Recall = \sum_c \frac{TP_c}{TP_c + FN_c}, c \in classes \tag{2}$$

$$Precision = \sum_c \frac{TP_c}{TP_c + FP_c}, c \in classes \tag{3}$$

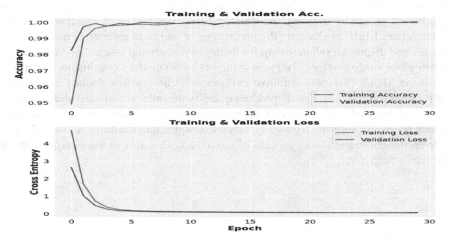

Fig. 4. Plot for training & validation accuracy and loss of proposed methodology

$$F1_{score} = 2 * \frac{precision * sensitivity}{presicion + sensitivity} \qquad (4)$$

```
              precision    recall  f1-score   support

           0       1.00      1.00      1.00       479
           1       1.00      1.00      1.00       519
           2       1.00      1.00      1.00       498
           3       1.00      1.00      1.00       488
           4       1.00      1.00      1.00       516

    accuracy                           1.00      2500
   macro avg       1.00      1.00      1.00      2500
weighted avg       1.00      1.00      1.00      2500
```

Fig. 5. Classification on test dataset of proposed methodology.

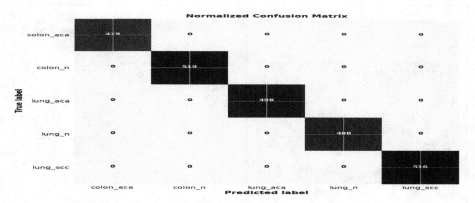

Fig. 6. Confusion matrix of proposed methodology.

Table 2. Images and their corresponding saliency map.

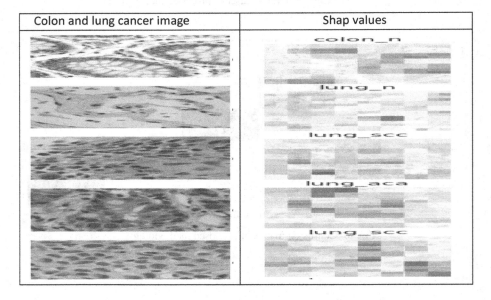

Table 3. Images and their corresponding shap values

5 Conclusion and Discussion

The histopathological image classification approach presented in this paper showcases its ability to achieve exceptional accuracy, especially when dealing with extensive datasets. The aim is for physicians to employ this method to aid in clinical diagnosis, highlighting its potential practical implementation in real-world medical settings. The trained model assesses its performance on a test dataset and generates a classification report. The paper showcases the model's flawless performance in precision, recollection, and F1-score for both colorectal and lung classes. The total accuracy is perfect, with a value of 1.00, signifying flawless categorization of the entire test dataset. The model demonstrates exceptional predictive capacity, as seen by the presented metrics. The study "Diagnosis of Pulmonary and Colonic Cancer" utilises Convolutional Neural Networks (CNN) and Contrast Limited Adaptive Histogram Equalisation (CLAHE) to achieve outstanding accuracy in automatically classifying cancer. This highlights the significant potential of deep learning in the field of medical imaging. Nevertheless, the study recognises several constraints, including the manual establishment of parameters for deep learning and a restricted emphasis on comprehensibility for studies that are not based on images.

References

1. Borkowski, A.A., Bui, M.M., Thomas, L.B., Wilson, C.P., DeLand, L.A., Mastorides, S.M.: Lung and colon cancer histopathological image dataset (lc25000). arXiv preprint arXiv:1912.12142 (2019)
2. Hamed, E.A.R., Salem, M.A.M., Badr, N.L., Tolba, M.F.: An efficient combination of convolutional neural network and LightGBM algorithm for lung cancer histopathology classification. Diagnostics **13**(15), 2469 (2023)
3. Martínez-Fernandez, E., Rojas-Valenzuela, I., Valenzuela, O., Rojas, I.: Computer aided classifier of colorectal cancer on histopatological whole slide images analyzing deep learning architecture parameters. Appl. Sci. **13**(7), 4594 (2023)
4. Tummala, S., Kadry, S., Nadeem, A., Rauf, H.T., Gul, N.: An explainable classification method based on complex scaling in histopathology images for lung and colon cancer. Diagnostics **13**(9), 1594 (2023)
5. Ahmed, M.S., Iqbal, K.N., Alam, M.G.R.: Interpretable lung cancer detection using explainable AI methods. In: 2023 International Conference for Advancement in Technology (ICONAT), pp. 1–6. IEEE (2023)
6. Hadiyoso, S., Aulia, S., Irawati, I.D.: Diagnosis of lung and colon cancer based on clinical pathology images using convolutional neural network and CLAHE framework. Int. J. Appl. Sci. Eng. **20**(1), 1–7 (2023)
7. Qasthari, B.L., Susanti, E., Sholeh, M.: Classification of lung and colon cancer histopathological images using convolutional neural network (CNN) method on a pre-trained models. Int. J. Appl. Sci. Smart Technol. **5**(1), 133–142 (2023)
8. Masud, M., Sikder, N., Nahid, A.A., Bairagi, A.K., AlZain, M.A.: A machine learning approach to diagnosing lung and colon cancer using a deep learning-based classification framework. Sensors **21**(3), 748 (2021)
9. Sakr, A.S., Soliman, N.F., Al-Gaashani, M.S., Pławiak, P., Ateya, A.A., Hammad, M.: An efficient deep learning approach for colon cancer detection. Appl. Sci. **12**(17), 8450 (2022)
10. Yahia Ibrahim, N., Talaat, A.S.: An enhancement technique to diagnose colon and lung cancer by using double CLAHE and deep learning. Int. J. Adv. Comput. Sci. Appl. **13**(8) (2022)
11. Guo, H., Kruger, U., Wang, G., Kalra, M.K., Yan, P.: Knowledge-based analysis for mortality prediction from CT images. IEEE J. Biomed. Health Inform. **24**(2), 457–464 (2019)
12. Nasser, I.M., Abu-Naser, S.S.: Lung cancer detection using artificial neural network. Int. J. Eng. Inf. Syst. (IJEAIS) **3**(3), 17–23 (2019)
13. Pang, S., Zhang, Y., Ding, M., Wang, X., Xie, X.: A deep model for lung cancer type identification by densely connected convolutional networks and adaptive boosting. IEEE Access **8**, 4799–4805 (2019)

14. Liu, L., Dou, Q., Chen, H., Qin, J., Heng, P.A.: Multi-task deep model with margin ranking loss for lung nodule analysis. IEEE Trans. Med. Imaging **39**(3), 718–728 (2019)
15. Shakeel, P.M., Burhanuddin, M.A., Desa, M.I.: Lung cancer detection from CT image using improved profuse clustering and deep learning instantaneously trained neural networks. Measurement **145**, 702–712 (2019)
16. Li, Y., Zhang, L., Chen, H., Yang, N.: Lung nodule detection with deep learning in 3D thoracic MR images. IEEE Access **7**, 37822–37832 (2019)
17. Chen, W., Wei, H., Peng, S., Sun, J., Qiao, X., Liu, B.: HSN: hybrid segmentation network for small cell lung cancer segmentation. IEEE Access **7**, 75591–75603 (2019)
18. Lakshmanaprabu, S.K., Mohanty, S.N., Shankar, K., Arunkumar, N., Ramirez, G.: Optimal deep learning model for classification of lung cancer on CT images. Futur. Gener. Comput. Syst.Comput. Syst. **92**, 374–382 (2019)

Deep Learning Approaches for Chest X-Ray Based Covid-19 Identification

R. Prabu[✉] and P. Dhinakar

Department of Electronics and Communication Engineering, Bharath Institute of Higher Education and Research, Chennai, Tamil Nadu, India
`prabu.6037@gmail.com, dhinakar.ece@bharathuniv.ac.in`

Abstract. Over 200 million cases of infection and 4 million fatalities have been attributed to the COVID-19 pandemic worldwide. Numerous people's lives and health are being significantly impacted globally by the pandemic-induced epidemic. The preservation of social well-being greatly depends on early identification of this illness. COVID-19 detection most commonly relies on RT-PCR testing, but it is not the only instrument capable of detecting COVID-19. One of the most advanced technologies on the market today, deep learning has been used to successfully treat a wide range of medical conditions. This work leads us to examine image-based techniques for coronavirus detection. People's respiratory systems are impacted by coronavirus. Chest radiography pictures are a valuable resource for diagnosing this condition. Chest radiographs of covid positive individuals showed distinct abnormalities, according to early study. Our Deep Learning Multi-layered networks identified covid positive and negative chest photos. A dataset of Coronavirus patients is used in the model under consideration, where the radiologist reports multilobar lesions.

Keywords: X-Ray · Covid-19 · Deep Learning · RT-PCR testing · World Health Organisation

1 Introduction

The term "novel" is commonly used to describe a new strain of a coronavirus, which is a hazardous virus family [1]. According to the World Health Organisation, coronavirus belongs to a large virus family that encompasses everything from common colds to serious disorders. COVID-19 has caused several times as many fatalities (>3M) as the combined deaths of MERS and SARS (about 1700) [3], but having a lower mortality rate (around 3% [2]) than SARS (10%) and MERS (35%). Through person-to-person contact, the SARS-CoV outbreak has spread to 26 countries globally [4]. Over 1600 patients in 27 countries were affected by the MERS-CoV pandemic in 2012, which led to over 600 fatalities, with Saudi Arabia accounting for 80% of those cases [5]. As a result of the current COVID-19 epidemic, which has spread around the globe, the World Health Organisation (WHO) was forced to designate the illness a pandemic on March 11, 2020. The global dynamics of business, economy, and society were impacted.

Governments have enforced social separation, flying limitations, and increased hygienic awareness. Still, COVID-19 is dispersing really quickly. Pneumonia, coughing, fever, and shortness of breath are typical coronavirus symptoms. Acute respiratory distress syndrome (ARDS) or total respiratory failure are severe instances of coronavirus infections that necessitate mechanical ventilation and admission to an intensive care unit. Severe illnesses, such as organ failure, especially renal failure or septic shocks, are more common in the elderly, in those with weakened immune systems, and in people with other chronic conditions [6]. One of the main concerns with the COVID-19 pandemic is its faster spreading pace, hence early detection of COVID-19 viral infection is essential [7] in order to stop the virus's spread. The recognised standard diagnostic technique is real-time polymerase chain reaction (RT-PCR) for the identification of viral nucleic acids [8]. However, the sensitivity and specificity of this test are not ideal, and many hyperendemic areas and nations are unable to quickly and adequately conduct RT-PCR testing for tens of thousands of suspected individuals. Other problems with RT-PCR include its discomfort, the high false-negative rate, the lack of swabs, the need for reagents, and the duration of time it takes to obtain results. In the light of these concerns, other diagnostic techniques have to be investigated. However, most of the suggested AI-based COVID-19 identification methods aim to differentiate COVID-19 from other bacterial or viral diseases or from standard X-rays. To the best of the author's knowledge, however, no research has been done in the literature to distinguish COVID-19 infection from the other two members of the COVID family, MERS and SARS. Because lung infections have overlapping patterns, medical professionals (MDs) find it challenging to differentiate between pictures from distinct CoV family members using CXR alone without the assistance of clinical data. As a result, examining how similar COVID family members appear to AI can yield insightful information that might aid in medical diagnosis [9]. It is widely acknowledged that RT-PCR is the gold standard of DNA testing; however, due to sample errors and a high time requirement, this test can be inaccurate and time-consuming. The sensitivity of these tests is insufficient for early detection of the virus. Thus, a quicker, more dependable, and more accurate diagnostic method is required. Both the respiratory organ and the upper respiratory tract get infected by the coronavirus. The characteristic characteristics of COVID-19 appear on radiography images as bilateral, subpleural, ill-defined opacities on ground glass. Even in afflicted individuals who are asymptomatic, abnormal lung radiography scans may be seen. Lesions that are symptomatic for one to three weeks can quickly develop into ground-glass opacities, increasing in approximately two weeks [10]. The CT results of the 114 patients who recovered from coronavirus-induced pneumonia revealed that 38 displayed fibrotic-like abnormalities six months after their recovery. These alterations were linked to advanced age, prolonged hospital admissions, arrhythmias, ventilator usage, and acute respiratory distress syndrome. A study by [11] showed how to use deep learning model ensembles that have been gradually pruned to identify COVID-19 pulmonary symptoms on chest X-rays. In this study, trimmed models with a weighted average accuracy of 99.01% significantly improve performance. In their investigation, they classified COVID-19 and normal chest X-ray pictures using deep learning algorithms. To extract deep features, we employed ResNet18, ResNet50, ResNet101, VGG16, and VGG19. At 94.7%, the deep features that were obtained using the ResNet50 model had the highest overall accuracy.

A pre-trained ResNet50 model was used in X. Gu et al.'s (2018) study. By comparing similarities between test pictures and the input image, the closest neighbour method was used. The model showed 89.7% accuracy after evaluation. According to [14], by utilising Inception V3 with transfer learning, chest X-ray radiographs might be utilised to diagnose coronavirus pneumonia in unwell individuals. It has a 98% classification accuracy. The primary goal of the presented study is to create a deep feature extraction tool for classification utilising a standard neural network. The method makes use of X-ray pictures to assist in the screening of COVID-19 samples.

2 Materials and Methods

One of the most important steps in resolving a research issue is computational thinking. The ability to think through an issue and use technology to solve it is crucial. First, we must determine whether the issue can be resolved. Before we can answer the problem or design an algorithm, we must first demand a more conceptual approach. Then, we would use the logical strategy and instruct the machine to do so using a programming language that is appropriate [15]. Consequently, we may believe that computational thinking entails drawing up a plan and then carrying it out. A deep learning multi-layered network is used in this study to categorize chest pictures as either positive or negative for COVID-19.Because chest X-rays are a less dangerous and more affordable alternative to CT scans because to their lower radiation doses, we utilise them to detect COVID-19. For our investigation, we employ X-ray scans since they have numerous benefits over CT. The chest x-rays of four individuals are shown in Fig. 1, with two of the infected patients showing multilobar involvement.

Sixty-five hundred of the chosen images—both positive and negative COVID-19 cases—are utilised for training, while the remaining fifteen hundred are used for model testing. There are 299,299 pixels in the photos, and they are saved as Portable Network Graphics (PNG). The aim of this study is to identify whether or not chest X-ray pictures are positive for COVID by employing cutting-edge algorithms. The chest X-rays of four participants, of which two are coronavirus-positive, are displayed in Fig. 1. Implementing multilayered deep learning technologies can help automate the diagnosis of patients who are infected and those who are not. More accurate results are produced by deep learning models, particularly when it comes to classifying images. From the input, we retrieved the features. After training the model, we assessed it to determine how well it was doing. For this study, we used the Google Colab GPU. Pixel dimensions of the input photos used are 299×299 pixels.The input layer is made up of 89,401 neurons in total. Activation of the Rectified Linear Units (ReLu) is set for the neurons. In this study, an Intel i3 CPU running Windows 10 was used to experimentally test the CNN model using Python.

2.1 Networks of Convolutional Neurons

Deep Learning (DL) uses a Convolutional Neural Network (CNN) as its basic model. These models don't require any user-defined characteristics because they are built to adapt from data. CNNs can be larger and more complex than Artificial Neural Networks (ANNs). CNNs are defined as the deeper network that emerges from adding more hidden

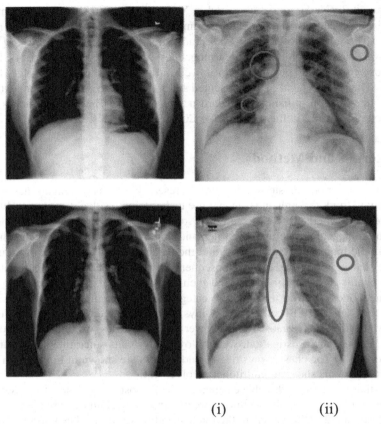

(i)   (ii)

Fig. 1. Images of four patients' chest X-rays (i) Normal (ii) Coronavirus-affected

layers to a neural network. In CNN design, 2-D filters are typically used. CNN has been a widely used technique, particularly for medical image processing and research [16].

2.2 Deep Learning (DL)

The fundamental architecture of a deep learning neural network is shown in Fig. 2. Additionally useful for analysing unstructured data is deep learning. An artificial neural network that can learn from enormous amounts of data is an example of deep learning - a subset of machine learning inspired by the similarities between the human brain and machine learning methods. Deep learning is facilitated by a neural network with several deep layers. Deep learning is particularly useful for analysing rich, complicated, multidimensional data, including images, videos, and audio. The model for categorising people infected with the coronavirus as first, second, third, and fourth convolution layers is shown in Fig. 3. In this model, the pooling layers are represented by P1, P2, P3, and P4.R1, R2, R3, and R 4 reflect the ReLu activation function.

2.3 Convolution Layer (CL)

Convolution is essential to the functioning of a network, as its name suggests. Since it does the majority of the computation, this layer is considered the network's core. Training kernels are the main focus of the layer's parameters. There are typically few spatial dimensions in these kernels, but they unroll deeply over all dimensions of data. By identifying the kernels that, for a particular set of input data, function best for the network, CNN models may be built.

The main part of a CL that is dynamically learnt during training is called a kernel. Figure 4 illustrates the creation of a 2D activation feature map that results from a convolution layer convolving the input data across the spatial dimensions of the data.

2.4 Pooling

Other layers in CNN besides convolution layers are called pooling layers (PL). The suggestion is that the convolution layer's outputs be fed into the network's pooling layers. Pooling layers work to reduce the data's dimensionality over time, which lowers the coefficients and the cost of the model accordingly.

2.5 Flattening Layer (FL)

As seen in Fig. 6, a flattening layer transforms mapped data into a one-dimensional array. The FL is a sophisticated downsampling approach that, while downsampling an $X \times Y$ feature map into a 1×1 array, finds the average of all of its components, preserving information about each feature map. The flattened layer provides input to the thick layer or the fully connected layer. Deep learning requires a lot of matrix operations, which GPUs can substantially parallelize and speed up. A GPU can have an unlimited number of cores, but a CPU typically has a set number. Two problems impede GPU's technical viability: Long training times and finite GPU memory, which is severely affected by growing network sizes. Cross entropy is a well-liked cost function for classification problems. For complicated issues, the Cross Entropy function might be utilised instead of the Mean Squared Error function. Gradient Descent is a simplification technique that improves DL network-based models by minimising the cost function or error. It is employed to adjust our model's parameters. The cost function's gradient is determined via the gradient descent procedure. A gradient can assess variations in the weight based on changes in the cost function. It's an error minimization technique that makes use of a gradient descent optimizer. It measures adaptive learning thresholds for different parameters by projecting the moment of gradients. The Adam optimizer aids in enhancing CNN's capacity for categorization. Adam keeps track of earlier facts or knowledge. In the same way that momentum maintains a continuously declining mean of past gradients, Adam maintains previous gradients. To train the DL network, the available data is divided into training and testing sets. A total of 6500 images are used in this study, with 5000 serving as training images and 1500 serving as testing images. For the CNN's training, a batch size of 64 and an epoch of 120 were used. The training data set is an input for learning, whereas the testing set is used to forecast the model. The batch size remains unaffected even after the input is flattened using the Flatten class. ReLU is preferred

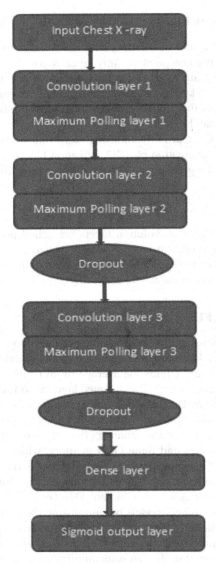

Fig. 2. Model architecture.

over other functions due to its ability to train network models more quickly. Only when the input is affirmative does the ReLU function output it. The ReLu activation function and the Dense class were used to compute the hidden layer. Since we only need binary outputs, the ReLu activation function is substituted with the Sigmoid function for the output. To calculate the cross entropy loss between real and predicted labels, use binary cross-entropy. In convolution neural networks, the hidden layers are a series of layers that pass through the input data. The multi-layered neural networks' primary operating

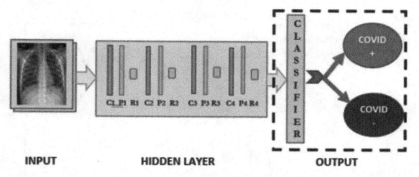

Fig. 3. DL-based categorization model

operations are depicted in Fig. 2. A feature detector, also known as a filter, is first convolved with the input picture in order to get matching feature maps. 3×3 filters have often produced better results. The following stage, known as pooling, involves gradually reducing the spatial size of the representation. This will facilitate computation and enable the machine to read the image. After every Convolution, Pooling comes next. The four map dimensions that we have utilised here are 32, 64, 64, and 128. We may state that the dimensional volume has expanded since 32 has grown to 64. A flattened signal is used by the Fully linked Layer (FC) to ensure that every response is linked to every neuron. One FC layer and four convolutional layers make up the weighted layers of the proposed DL model. Using the CNN Model, 1,735,937 parameters were recovered during the course of a three-hour training session. ReLU is the name of the common activation function used in picture classification. The FC layer is sent into the Sigmoid classifier, which utilises it as an input to assess if the patient is covid-positive. SIgmoid functions are a suitable output unit for the binary classification issue. The label for the output is either covid positive or covid negative, depending on the features that the network model evaluated on the input image.

3 Results and Discussions

The DL model is trained by splitting the available data into training and testing sets. The training dataset is used to train the DL model, while the testing dataset is utilised for validation processes. Initially, the model was trained across 50 epochs. Figures 4 and 5 show plots of the accuracy and cost functions (for 50 epochs). A total of 6500 chest X-ray radiograph images are used in this study; 5000 are used for training and 1500 are used for testing. Rescaling the X-ray images is part of the pre-processing step. Figures 4 and 5 show plots of the accuracy and cost functions, respectively. The results showed that the training accuracy was 90% and the validation accuracy was 89%. Figure 6 illustrates the model's predictive ability by showing an improvement in validation accuracy over training accuracy. The model has a validation accuracy of 94% and a training accuracy of 96% across 120 epochs. The training cost function is for 50 and 120 epochs, respectively, as Figs. 5 and 7 demonstrate. The performance of the classifier is estimated.

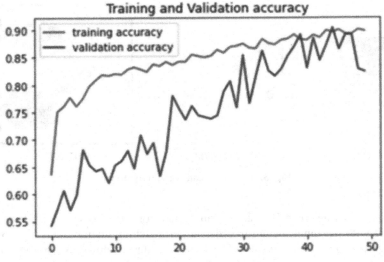

Fig. 4. A 50-epoch accuracy test

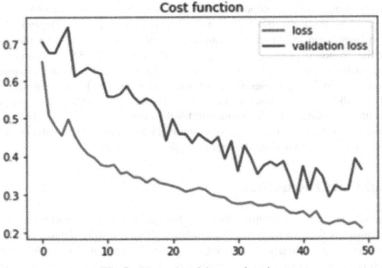

Fig. 5. 50 epochs of the cost function

Table 2 presents a comparison of the findings, showing that our model can categorise the randomly picked photos with a classification accuracy of around 94%, and just one real positive or incorrect prediction. There is an overall classification accuracy achieved by the suggested DL model. In order to improve the network, 64 was selected as the batch size. After the network model was verified, its accuracy increased to 94% from 96% when it was first trained.

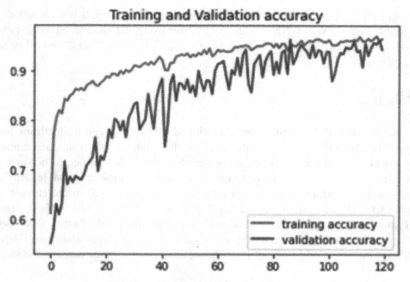

Fig. 6. An analysis of 120 epochs of accuracy

This outcome shows that the model can correctly detect patients who test positive for COVID-19. Table 2 contrasts different approaches with the outcome. CNN is also used to compare the outcomes with the suggested effort. The most widely used performance measures in deep learning are recall, accuracy, precision, and F1 score. [17] obtained an F1 score of 83.2% in their work on the creation and detection of COVID-19 infection

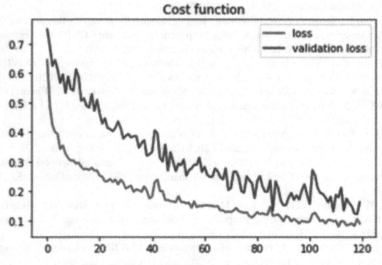

Fig. 7. Plots of the 120 epoch cost functions

maps from chest X-ray images. The recommended effort is what has caused the F1 score to increase to 90%. CNN is a useful tool for using the recommended computer-aided design approach to distinguish between Covid-19 positive and Covid-19 negative circumstances.

4 Conclusion

Numerous research organisations have developed different screening facilities for the purpose of identifying COVID-19 individuals, as the pandemic has resulted in numerous deaths worldwide. Of these, all of the answers have been determined to be somewhat inaccurate. But it's critical to identify this fatal illness as soon as possible. An X-ray picture-based method and a Deep Learning technique were used in the current work to predict COVID-19s. An algorithm developed by CNN has shown 94% accuracy in identifying the COVID-19 class and 90% F1-score for this class of viruses. X-rays are necessary for lung infection screening. The dependability of this modality will increase with advancements in AI (artificial intelligence)-based early-stage detection systems.

Conflicts of Interest. There are no conflicts of interest, according to the authors.

References

1. Huang, C., et al.: Clinical features of patients infected with 2019 novel coronavirus in Wuhan. China, The Lancet **395**(10223), 497–506 (2020)
2. Zhao, W., Zhong, Z., Xie, X., Yu, Q., Liu, J.: Relation between chest CT findings and clinical conditions of coronavirus disease (COVID-19) pneumonia a multicenter study. American Journal of Roentgenology **214**(5), 1072–1077 (2020)
3. Fu, S., Fu, X., Song, Y., Li, M., Pan, P.H., Tang, T., Zhang, C., Jiang, T., Tan, D., Fan, X., Sha, X.: Virologic and clinical characteristics for prognosis of severe COVID-19: a retrospective observational study in Wuhan, China. medRxiv **395**(1), 507 (2020)
4. Dong, D., et al.: The role of imaging in the detection and management of COVID-19: A review. IEEE Rev. Biomed. Eng. (2020). https://doi.org/10.1109/RBME.2020.2990959
5. Chadaga, K., Prabhu, S., Vivekananda, B.K., Niranjana, S., Umakanth, S., Pham, D.T.: Battling COVID-19 using machine learning: A review. Cogent Engineering **8**(1), 1958666 (2021). https://doi.org/10.1080/23311916.2021.1958666
6. Zhang, J., Xie, Y., Li, Y., Shen, C., Xia, Y.: COVID-19 screening on chest Xray images using deep learning based anomaly detection, (2020). arXiv preprint arXiv:2003.12338
7. Lu, C.-Y., Arcega Rustia, D.J., Lin, T.-T.: Generative adversarial network based image augmentation for insect pest classification enhancement. IFAC-PapersOnLine **52**(30), 1–5 (2019)
8. Saad, W., Shalaby, W.A., Shokair, M., Abd El-Samie, F., Dessouky, M., Abdellatef, E.: COVID-19 classification using deep feature concatenation technique. J. Ambient. Intell. Humaniz. Comput. **13**(4), 1–19 (2021)
9. Ozkaya, U., Ozturk, S., Barstugan, M.: Coronavirus (COVID19) classification using deep features fusion and ranking technique. (2020). arXiv preprint arXiv:2004.03698
10. Kermany, D.S., et al.: Identifying medical diagnoses and treatable diseases by image-based deep learning. Cell **172**(5), 1122–1131 (2018). https://doi.org/10.1016/j.cell.2018.02.010

11. Apostolopoulos, I.D., Aznaouridis, S.I., Tzani, M.A.: Extracting possibly representative COVID-19 biomarkers from X-ray images with deep learning approach and image data related to pulmonary diseases. Zhonghua yi xue gong cheng xue kan **40**(3), 462–469 (2020)
12. Khandekar, R., Shastry, P., Jaishankar, S., Faust, O., Sampathila, N.: Automated blast cell detection for acute lymphoblastic leukemia diagnosis. Biomed. Signal Process. Control **68**(102690), 1746–8094 (2021). https://doi.org/10.1016/j.bspc.2021.102690
13. Gu, X., Pan, L., Liang, H., Yang, R.: Classification of bacterial and viral childhood pneumonia using deep learning in chest radiography, In: Proceedings of the 3rd International Conference on Multimedia and Image Processing, pp. 88–93 (2018)
14. Islam, M.Z., Islam, M.M., Asraf, A.: A combined deep CNN-LSTM network for the detection of novel coronavirus (COVID-19) using X-ray images. Inform. Med. Unlocked **20**, 100412 (2020)
15. Parikh, R., Mathai, A., Parikh, S., Sekhar, G.C., Thomas, R.: Understanding and using sensitivity, specificity and predictive values. Indian J. Ophthalmol. **56**(1), 45–50 (2008)
16. Gholamrezanezhad, A.: Coronavirus disease 2019 (COVID-19): A systematic review of imaging findings in 919 patients. American Journal of Roentgenology **215**(1), 87–93 (2020). https://doi.org/10.2214/AJR.20.23034
17. Bellemo, V., Burlina, P., Yong, L., Wong, T. Y., Ting, D.S.W.: Generative adversarial networks (GANs) for retinal fundus image synthesis. In: Computer Vision – ACCV 2018 Workshops. ACCV 2018. Lecture Notes in Computer Science (2018)

Pattern and Speech Recognition

Bi-LSTM Based Speech Emotion Recognition

Pagidirayi Anil Kumar[✉] and B. Anuradha

Department of ECE, S.V. University College of Engineering, Tirupati, Andhra Pradesh, India
anilkumar.pgke@gmail.com, anubhuma@yahoo.com

Abstract. Using audio files, Mel Frequency Cepstral Coefficients (MFCCs), and obtaining models features, this paper proposes the two-stage method for emotion recognition in speech using deep learning's Bidirectional-Long Short-Term Memory (Bi-LSTM) approach.. The suggested novel feature extraction method is used to build a Bi-LSTM that sequentially processes input sequences in both directions. This bidirectional processing enables the network to acquire information from both the past and future contexts, making it well-suited for tasks that need a thorough grasp of sequential data that contributes to emotion recognition. Including anger, happiness, sadness, disgust, fear, and neutrality from audio signals. Training and test datasets are analysed in Python Colab to produce good results compared to other deep learning techniques. The CREMA-D and SAVEE datasets are used to recognize precision, sensitivity (true positive rates) of uttered emotions, and spectrograms of human emotions.

Keywords: SER · Bi-LSTM · MFCC · Precision · True Positive Rates

1 Introduction

The vocal signal is the mode of engagement and communication that is most commonly used and well understood. Because of this, researchers are looking into ways to facilitate efficient and speedy communication between humans and machines. Around the middle of the 1950s, a number of studies were conducted on audio signals recognition, this refers to the process of transforming the auditory representation of a human voice into a sequence of individual words.. In spite of the enormous progress that has been made in speech recognition, natural communication between humans and machines is not something that is likely to occur very soon because machines are unable to comprehend the emotional state of the person speaking. Speech recognition has made fresh opportunities available in this domain. Speech recognition is a technique for extracting useful semantic information from audio signals and increasing the effectiveness of speech identification techniques. Speech recognition is primarily concerned with determining the speaker's emotional state from their speech, whereas voice emotion recognition is primarily concerned with improving the efficiency of voice recognition systems [1].

In order to recognize a message, speaker, and language, as well as to process and network data, elementary units are essential to any conventional voice recognition system. These elements are all necessary for the system to function properly. An enormous amount of study has gone into developing ways for translating human speech into text in the earlier several years. Understanding and eliciting specific feelings are the two primary goals of research in the field of SER, this is a field of research that is still in its initial stages. Experts are working hard to improve speech recognition technology so that robots and people can converse more effectively; nevertheless, the inability of machines to distinguish between human and machine voices remains a significant obstacle.

Many possible applications for speech emotion recognition exist, including the following: (1) man-machine relations in realistic circumstances, such as online videos; (2) computer film and educational video recognition tools; (3) vehicle on-board system applications that send a cryptic message about the driver's psychological state to the vehicle's automated tool; (4) investigative device for a specialist to provide therapy to the viruses; and (5) utilized for the purpose of computerized translation systems that are fluent in the language of the speaker [2].

Anger, joy, sadness, disgust, fear, and neutral are the categories used to classify people's emotions when they speak. A strong SER system will include voice signal spectrum harmony, improved voice quality classifications, and efficient identification capabilities. The vast widely held of research efforts that are related to SER are concentrated on locating the functional combination of the aforementioned attributes that is the most efficient and successful in preserving the emotional components of a speech signal. This particular study the results of SER make it abundantly evident that the feature synthesis, which served as the combinational approach for precise information categorization, was used. However, the issue is that the computing cost of the classifier and the classification accuracy of SER when utilising the feature fusion method are directly correlated with one another. This is due to the fact that while some characteristics are more useful than others, others may not be useful for emotion recognition altogether[3]. Specifically, this is owing to the fact that some of the features make a higher contribution than others.

In a manner analogous to that of machine learning, SER has just begun to benefit from the increased efficacy offered by the tools provided by deep learning. The development of SER system made use of an extensive selection of pre-processing and precise feature engineering methods, as well as Support Vector Machines (SVM), Hidden Markov Models (HMM), and Gaussian Mixture Models (GMM). The majority of the recent research concentrates on deep learning as a means of producing superior results. Different datasets are required in order to construct strategies for machine learning and to correctly categorise feelings. Natural datasets, generated datasets, and simulated datasets are the primary types of training datasets that can be used for SER applications. The accessible video and audio is utilised, regardless of whether it was broadcast on television or uploaded on the internet, so that we may retrieve the raw data sets from the source. In addition, the obtained databases that are kept at call centres are another option that can be examined for natural datasets. The production of semi-natural datasets begins with the writing of a scenario and continues with the hiring of voice actors with professional experience. Simulated datasets, which fall under the third category and are by far the most common, appear to be artificial. When compared to actors reading the identical lines, those providing the voiceovers create a variety of distinct emotions. The conventional approach to speech recognition makes use of a SER equipped with cutting-edge methods such as HMM, generalised HMM, and SVMs. A lot of highlighted innovation went into these methods, and it was common practice to reorganize the method's architecture if the featured applications were modified. On the other hand, the methods and tools of deep learning offer answers that can also be applied to the creation of SER. The application of these methods to differentiate between emotions and speech has been the subject of a significant amount of discussion and research. Recent SER research has used these strategies as part of a mix of deep learning techniques to address the problem at hand. Methods such as generative adversarial designs, autoencoders, Bi-LSTM networks, and advancements in recurrent neural networks (RNN) are examples of these techniques[4].

The flow of order for the set of work has been divided below. The work related to SER is illustrated in Section II. Procedure and suggested tasks are detailed in Section III. The results and tables of the simulation are presented in Section IV. Section V contains the conclusive part.

2 Related Work

Research proposes a Bi-LSTM based SER system using MFCC and GTCC features on Berlin EMO-DB database. Even with a 60% training rate, the system manages to attains good accuracy. The Berlin EMO-DB dataset is used for simulations, which consists of seven speech classes. Features from GTCC and MFCC, as well as their versions, are used to train the Bi-LSTM classifier. Overall, the Bi-LSTM classifier shows good improvement for speech emotion recognition [5].

The author enhanced the accuracy of speech emotion recognition utilizing an S-kNN classifier with the MFCC feature extraction approach, which is part of a machine learning system used in SER. It classifies different emotions with reference to neutral speech. The system utilizes CSV files containing datasets sourced from RAVDESS, CREMA-D, TESS, and SAVEE databases. The accuracy rates for female and male speakers are 92.4% and 86.8% respectively, with higher PPV rates compared to other classifiers. When compared to other classifiers, the method also reveals that it requires less training time. To further enhance accuracy and training time for speech emotion recognition, it is possible to combine many classifiers or feature extraction methods [6].

This research presents a two-stage deep learning-based LSTM technique for recognizing voice emotions from audio recordings utilizing MFCC. Next, a network is trained to recognize emotions including anger, sadness, disgust, fear, and neutral using the proposed unique feature extraction approach. The Python Colab platform is used to analyze training and testing datasets in order to produce better results than existing deep learning techniques. Using the CREMA-D and SAVEE datasets, we can identify human emotion spectrograms, true positive rates of uttered emotions, and precision [7].

Aims to understand certain characteristics of emotions in order to detect the pitch of voice. Two 1D CNN LSTM networks and one 2D CNN LSTM network were constructed to learn local and systemic emotion-related features from voice and log Mel spectrums. Two networks with the same architecture—one LSTM layer and four local feature learning blocks (LFLBs)—are in fact one and the same. Trained both for local and hierarchical correlation extraction, the LFLB primarily uses a max-pooling layer and a single convolutional layer [8].

The author proposed review on deep learning techniques for SER. It is a summary of various studies and their future directions. With limitations deep learning techniques are computationally complex. Combination of multiple classifiers may improve emotion detection. Limitations of the methods are mentioned but not specified clearly [9].

Emotion representation in speech is important for natural sounding synthesised speech. Various methods like Gaussian mixture models, HMMs, SVMs, CNNs, LSTMs have been used for SER. Simulation using Deep Neural Networks (DNNs) has shown favourable outcomes in the area of SER. Attention mechanisms have gained popularity in speech emotion recognition. Multi-modal approaches combining audio, visual, and textual attributes have been explored. Research on attention processes in emotion identification from spoken language has been limited. An examination of attention-based neural networks as a tool for emotion identification in spoken language is given in this research. Considering the constraints, it is not possible to verify the efficiency of rhythmic characteristics without considering spectral aspects. The dataset utilized is a synthesis of many datasets [10].

The present study is an example of a systematic review that takes a Machine Learning approach to studying SER. Focusing to the three main components of SER implementation: analysing information, selecting features, and categorization. Speaker-Independent experiments' poor precision for classification is one of the problems and solutions covered in this study of SER. There has been not much discussion about the difficulties and current best practices in SER execution, and Speaker-Independent experiments have a low categorization precision [11].

3 Proposed Work

Emotion recognition and computer-human interaction (or SER) is the broad definition of this technology. Multiple methods of speech signal analysis are combined in SER to help determine the underlying emotional state of a speaker's voice. Numerous models exist for analysing voice data, making predictions, and spotting latent feelings. This study makes use of RNN applications based on LSTM. Sequential data, like as audio files, can only be properly investigated and evaluated with the aid of the appropriate model. Therefore, the sequential LSTM model is used to actualize this technique. The primary objectives are to develop a system capable of identifying innate emotional content in voice signals, to enhance the precision of predictions and overcome obstacles caused by fading gradients in RNN.

3.1 Datasets

There are 7,442 uncut videos with 91 different actors in the Crowd-sourced Emotional Multimodal Performers Dataset (CREMA-D). The performers who took part in these recordings ranged in age from 20 to 74, were from a wide range of ethnic and cultural backgrounds (including African American, Asian, Caucasian, Hispanic, and Unspecified), and represented a wide range of sexes. The 12 possible statements are presented by the performers. Six distinct emotions (anger, disgust, fear, joy, neutral, and sadness) and four varied intensity levels (minimum, Medium, High, and Unspecified) were used to express the remarks. We started talking about how we felt and how intense those feelings were while seeing the video, listening to the music, and taking in the whole multimedia experience.

A dataset capturing Surrey Audio-Visual Expressed Emotion (SAVEE) was utilized in the development of an automated emotion recognition system. There are 480 words total, spoken by 4 male celebrities enacting 7 different emotions. Examples of phrases used in the TIMIT corpus were adapted to be grammatically consistent with each mood. The research was conducted in a multimedia lab with state-of-the-art audiovisual tools, and the results were examined and sorted. Ten participants watched the videos and listened to the audio files to ensure that everything was working properly. The auditory, optical, and cinematic techniques' classification methods were created with similarities in mind, utilizing methods and the components., yielding 61%, 65%, and 84% speaker-independent recognition rates, respectively [17]. Figure 1 displays a bar chart with the different emotions found in the CREMA-D and SAVEE datasets.

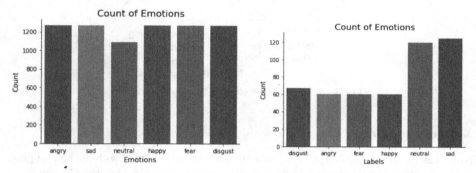

Fig.1. Number of emotions in the CREMA-D and SAVEE datasets.

In the method, after gathering raw speech data from databases, the initial stage is to pre-process the data. The flow diagram in Fig. 2 shows how the Bi-LSTM deep learning network and the MFCC feature extraction algorithm create layered data in a sequential order. The next step is to train and evaluate this data to accurately distinguish between various speech emotions. These emotions include disgust, fear, as well as joyful, sad, and neutral.

The pre-processing unit receives data about speech from the database. The next section describes the actions depicted in the block diagram.

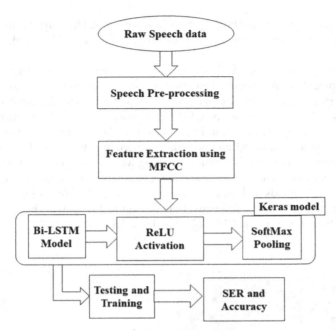

Fig. 2. Schematic flow chart of Proposed Method

3.2 Pre-processing

Noise can be reduced and high-frequency content can be amplified with the help of pre-emphasis filters like hamming windowing. The formula represented as Eq. (1), equation of filter.

$$z(t) = h(n) - \propto h(n-1) \tag{1}$$

where $\propto$ is the filter value that is generated by MFCC and has the value of 0.97 [13].

3.3 Feature Extraction

Feature extraction is essential for any machine learning model. A better trained model may result from careful feature selection, while training would be significantly hampered by erroneous feature selection. We used the MFCC to extract features.

Mel-Frequency Cepstral Coefficients (MFCC). The steps involved in the MFCC feature extraction approach are as follows: first, a Mel scale frequency range setup; second, signal application of the window function; third, discrete Fourier transform (DFT) derived from log magnitude; and lastly, DCT inversion [12]. In order to gain the features of MFCC, the following steps, which are outlined in the parts that follow, must first be completed (Fig. 3).

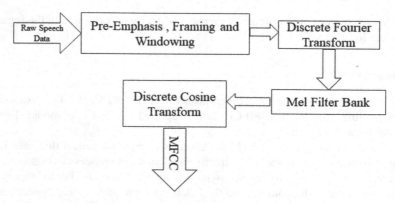

Fig. 3. MFCC Feature Extraction ([6], p. 273)

An audio signal's higher-frequency components are emphasized by the pre-emphasis filtering process. It strives to achieve harmony across the full spectrum of audible tones. When spoken sounds are included, the slope of the source is approximately 12 decibels per octave. When pre-emphasis is used, certain glottal effects produced by the vocal tract's features are removed. The duration of each frame ranges from 20 ms to 30 ms, and the frames that make up the vocal signal. M (MN) successive frames are used to differentiate the following frames from N samples collected from the voice stream. The default values for M and N are 100 and 256 respectively. Since speech is a signal that changes over time, it must be appropriately framed for quick analysis however, its properties are, for the most part, consistent. As a direct consequence of this, a spectrum analysis can only be carried out for a certain amount of time.

To keep the signal uniform across time, a hamming window function is used to each successive frame. The windowing mechanism helps mitigate this erratic behaviour at minimise distortion. While recording, a hamming window is applied at the beginning and finish of the voice sample.

$$W[n] = P[n] * Q[n] \tag{2}$$

Q[n] is hamming window.

A frequency domain to time domain translation is referred to as an FFT. The Fast Fourier Transform (FFT) approach is utilised in order to ascertain the magnitude frequency response of each frame. The Fast Fourier Transform (FFT) produces a spectrum, which is also known as a periodogram in some contexts. In order to get a smooth spectrum, twenty triangle bandpass filters are multiplied by the frequency response of magnitude. In addition, the ranges of the characteristics that are being investigated are somewhat restricted. The consistency of the vocal tract makes it possible for the energy levels of different bands to commonly coincide with one another. A grouping of the improvised Mel frequency coefficients are input into the DCT in order to obtain the cepstral coefficients [12].

$$mel(f) = 2595 \log_{10} 1 + \frac{f}{700} \tag{3}$$

The Mel-frequency, a non-linearity metric, is directly proportional to the logarithm of the linear frequency f.

3.4 Keras Model

In order to construct a network of hidden layers with Bi-LSTM, ReLU activation, and SoftMax pooling, the obtained MFCC data is applied to the Keras model. Figure 4 depicts the Keras LSTM architecture.

Bidirectional LSTM, or Bi-LSTM for short, is a sequence model that uses LSTM layers: one to process data in the front direction and another to process information in the reverse direction. It is typically applied to jobs using NLP. The idea behind this method is that the model can better comprehend the link between sequences (e.g., understanding the words that precede and follow in a phrase) by processing input in both directions.

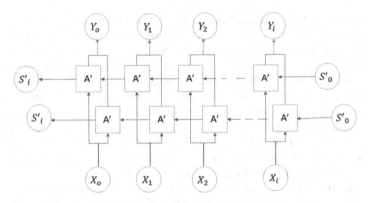

Fig. 4. Architecture of Keras Bi-LSTM

When compared to unidirectional LSTM, the bidirectional LSTM aids the machine in comprehending this relationship more fully. Because of this feature, Bi-LSTM is an appropriate architecture for machine translation, sentiment analysis, and text classification applications.

$$a_t = a_t^f + a_t^b \quad (4)$$

a_t = Network final probability vector.

a_t^f = The probability vector from the forward LSTM network.

a_t^b = The backward LSTM network's probability vector.

The Bi-LSTM layer design is shown in the above picture, where A and A' are LSTM nodes, X_i is the input token, and Y_i is the output token. The combination of LSTM nodes A and the end result of the Yi algorithm is A'. Bidirectional learning differs from unidirectional learning in that it runs inputs in two directions, one from the past to the future and the other from the future to the past. This allows you to remember information from both the past and the future simultaneously by combining the two hidden states.

3.5 Validation of the Model

Data for training (train X), data for target (train y), validation of testing data, and epoch numbers are all sent into the fit() method, which in turn trains the network model. For cross-validation, we use the X and y tests on a subset of the dataset. The data is processed repeatedly by the model for the given number of epochs. The system can be optimized by running many epochs, however there is a limit beyond which the custom-built system for each epoch will no longer function. Thirty epochs will be used in the implementation of the suggested model network [21]. The training of the model can now begin. The "fit" function is utilised during training for the intended model.

```
Model: "sequential"
_____
Layer (type)                    Output Shape              Param #
=================================================================
bidirectional (Bidirectiona     (None, None, 256)         145408
l)

dropout (Dropout)               (None, None, 256)         0

bidirectional_1 (Bidirectio     (None, 256)               394240
nal)

dropout_1 (Dropout)             (None, 256)               0

dense (Dense)                   (None, 64)                16448

dropout_2 (Dropout)             (None, 64)                0

dense_1 (Dense)                 (None, 10)                650

=================================================================
Total params: 556,746
Trainable params: 556,746
Non-trainable params: 0
_____
```

Fig. 5. Parameter table.

The model's architecture in Fig. 5, which shows the total number of inputs (both training and testing), the format of the layer outputs, and the types of layers utilized. Model evaluation is a crucial step in the model building process. It aids in selecting the optimal model for defining the provided data and predicting the optimal future performance of the selected model. The problem of overfitting can be avoided through analysis of prediction accuracy using a test set. The assessment's focus is on the reliability and precision of forecasted future data. In the findings section, we talk about the data we got from our experiments.

4 Results and Discussions

Using a trained result of Bi-LSTM model, we are able to recognise a variety of emotions with high precision. Experimental results show that using 30 epochs and a dropout rate of 0.3 improves the accuracy of CREMA-D and SAVEE's voice data. Figure 6 displays the wave graphs and spectrums for each emotion of datasets.

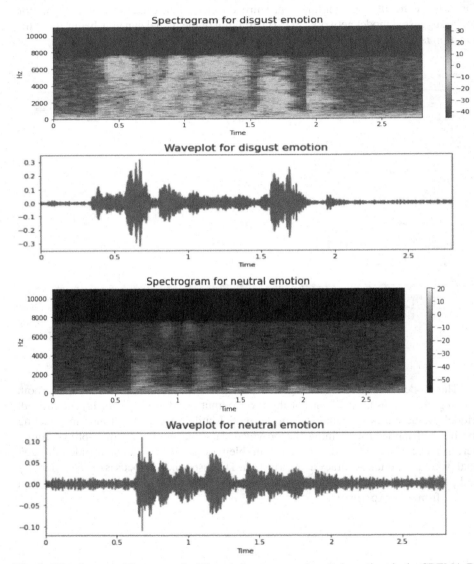

Fig. 6. Waveforms and Spectrums for Disgust, fear, angry and neutral emotions in the CREMA-D and SAVEE dataset.

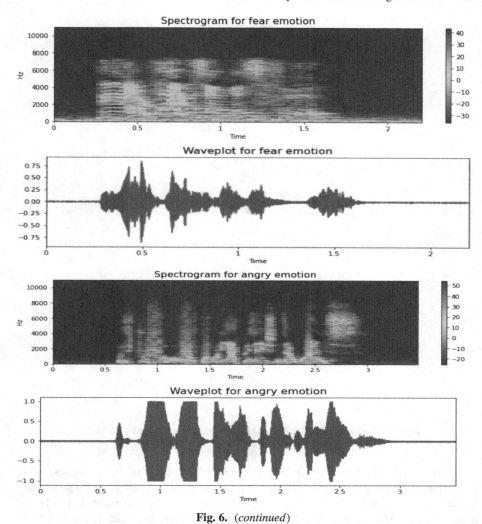

Fig. 6. (*continued*)

All of the emotions portrayed in the two databases' speech samples are shown in Fig. 6 wave graphs and spectra, which include disgust, fear, happiness, anger, sadness, and neutrality.

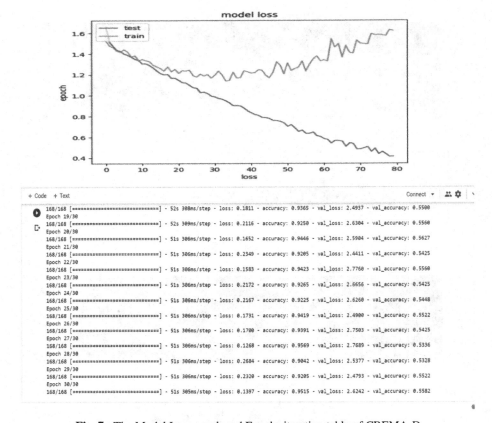

Fig. 7. The Model Loss graph and Epochs iteration table of CREMA-D

Figures 7, 8 and 9 displays the total number of emotions in the databases, while Fig. 10 total count of emotions and Figs. 11 and 12 shows the final precision attained by each emotion in CREAMA-D and SAVEE Datasets. The precision was obtained using the bi-LSTM (proposed) approach. Compared to other approaches, the Bi-LSTM methodology improves them in terms of efficiency with true positive rate.

The above Tables 1, 2, 3 and 4 represents the true positive rates and precision of obtained from proposed method with CREAMA-D and SAVEE datasets with other methods.

Figures 13 and 14 show graphs of the novel technique's accuracy and true positive rates of proposed method with comparison to other methods plotted from Tables 1, 2, 3 and 4.

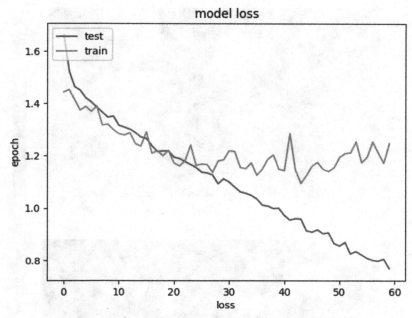

Fig. 8. The Loss graph for SAVEE

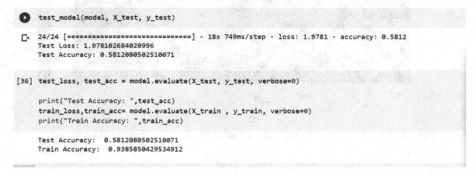

Fig. 9. Epochs iteration of SAVEE Databases using proposed method.

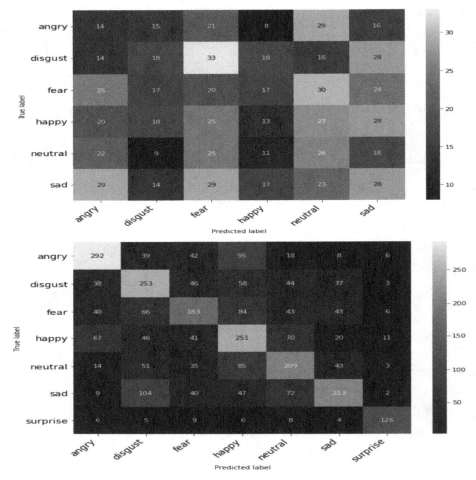

Fig. 10. Count of Emotions in confusion matrices for SAVEE and CREAMA-D

	Disgust	Angry	Fear	Happy	Neutral	Sad
Disgust	95.75	0	0	2.41	1.01	0.83
Angry	0	92.00	4.55	0	0	3.45
Fear	1.84	0	94.87	1.82	1.47	0
Happy	2.72	0	0	94.28	3.0	0
Neutral	0	1.92	0	0	93.58	4.50
Sad	0	0	4.88	0	0	95.12
	Disgust	Angry	Fear	Happy	Neutral	Sad
Disgust	**96.15%**	1.00%	2.80%	0.00%	0.05%	0.00%
Angry	2.50%	**92.87%**	3.18%	1.10%	0.00%	0.35%
Fear	3.80%	3.60%	**94.27%**	0.40%	0.10%	0.20%
Happy	0.37%	2.45%	1.40%	**95.08%**	0.50%	0.20%
Neutral	0.00%	0.80%	0.90%	0.67%	**94.18%**	3.45%
Sad	1.15%	0.87%	1.58%	0.28%	0.20%	**95.92%**
PPV	96.15%	92.87%	94.27%	95.08%	94.18%	95.92%
FDR	3.85%	7.13%	5.73%	4.92%	5.82%	4.08%

Fig. 11. Confusion matrices and True Positive Rates obtained by the proposed method for CREMA-D

	Disgust	Angry	Fear	Happy	Neutral	Sad
Disgust	**95.75**	0	0	2.41	1.01	0.83
Angry	0	**92.00**	4.55	0	0	3.45
Fear	1.84	0	**94.87**	1.82	1.47	0
Happy	2.72	0	0	**94.28**	3.0	0
Neutral	0	1.92	0	0	**93.58**	4.50
Sad	0	0	4.88	0	0	**95.12**

	Disgust	Angry	Fear	Happy	Neutral	Sad
Disgust	**96.15%**	1.00%	2.80%	0.00%	0.05%	0.00%
Angry	2.50%	**92.87%**	3.18%	1.10%	0.00%	0.35%
Fear	3.80%	3.60%	**94.27%**	0.40%	0.10%	0.20%
Happy	0.37%	2.45%	1.40%	**95.08%**	0.50%	0.20%
Neutral	0.00%	0.80%	0.90%	0.67%	**94.18%**	3.45%
Sad	1.15%	0.87%	1.58%	0.28%	0.20%	**95.92%**
PPV	**96.15%**	**92.87%**	**94.27%**	**95.08%**	**94.18%**	**95.92%**
FDR	3.85%	7.13%	5.73%	4.92%	5.82%	4.08%

Fig. 12. Confusion matrices and TPR obtained by the Bi-LSTM for and SAVEE datasets.

Table 1. Efficiency of the proposed work for CREAMA-D and SAVEE Datasets.

S.No	Technique	Accuracy (Efficiency)
1.	Proposed	**88.57%**
2.	Deep Learning methods [15]	74.07%
3.	1D & 2D CNN [14]	52.14%
4.	CNN [16]	85.2%

Table 2. True Positive Rate Efficiency of the proposed work for CREAMA-D

S.No	Technique	Accuracy
1.	Proposed	**89.78%**
2.	Deep Learning methods [15]	72.83%
3.	1D & 2D CNN [14]	62.07%
4.	CNN [16]	81.9%

Table 3. Efficiency of the proposed work for SAVEE

S.No	Technique	Accuracy
1.	Proposed	**93.85%**
2.	1D & 2D CNN [14]	89.16%
3.	3D CNN [13]	86.18%
4.	CNN [16]	73.60%

Table 4. True Positive Rate Efficiency of the proposed work for SAVEE

S.No	Technique	Accuracy
1.	Proposed	**94.67%**
2.	1D & 2D CNN [14]	89.16%
3.	3D CNN [13]	86.18%
4.	CNN [16]	73.60%

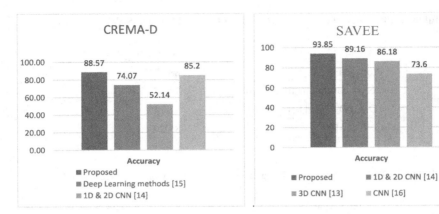

Fig. 13. Evaluation of precision obtained for Bi-LSTM.

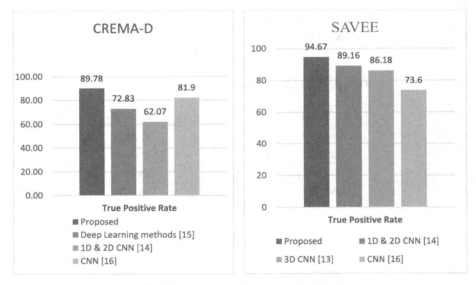

Fig.14. Evaluation of TPR for Bi-LSTM.

5 Conclusion

In the SER system utilizing MFCC feature extraction, the proposed DNN based on a Bi-LSTM architecture effectively conveys a range of emotions, including anger, disgust, fear, gladness, and sadness, in comparison to neutral speech. This study employs the Keras model-driven Bi-LSTM approach along with the MFCC feature extraction technique, utilizing datasets from the CREMA-D and SAVEE audio clip databases. The Bi-LSTM network exhibits outstanding performance, achieving an efficiency of 88.5% for CREMA-D and a positive predictive value (PPV) rate of 93.85% for SAVEE, better the results of previous methodologies. Consequently, the Bi-LSTM network model demonstrates superior precision in classifying emotional states in speech compared to alternative classifiers but consumes time for iterations of epochs. The integration of feature extraction methods with classifiers enhances both accuracy and training efficiency in detecting speech emotions through Artificial Intelligence (AI).

References

1. El Ayadi, M., Kamel, M.S., Karray, F.: Survey on speech emotion recognition: features, classification schemes, and databases. Pattern Recogn. **44**, 572–587 (2011)
2. Swain, M., Routray, A., Kabisatpathy, P.: Databases, features and classifiers for speech emotion recognition. Int. J. Speech Technol. **21**, 93–120 (2018)
3. Bandela, S.R., Kishore Kumar, T.: Speech emotion recognition using unsupervised feature selection algorithms. Radio Eng. **29**(2), 353–364 (2020)
4. Abbaschian, B.J., Sierra-Sosa, D., Elmaghraby, A.: Deep learning techniques for speech emotion recognition, from databases to models. Sensors **21**, 1249 (2021)
5. Subbarao, M.V., Terlapu, S.K., Chowdary, P.S.R.: Emotion recognition using BiLSTM classifier. In: IEEE 2022 International Conference on Computing, Communication and Power Technology (IC3P), June 2022. https://doi.org/10.1109/IC3P52835.2022.00048
6. Pagidirayi, A.K., Bhuma, A.: Speech emotion recognition using machine learning techniques. Revue d'Intelligence Artificielle **36**(2), 271–278 (2022). https://doi.org/10.18280/ria.360211
7. Pagidirayi, A.K., Anuradha, B.: An efficient speech emotion recognition using LSTM model. NeuroQuantology **21**(1), 117–127 (2023). Bornova Izmir. https://doi.org/10.48047/nq.2023.21.01.NQ20007
8. Singh, J., Saheer, L.B., Faust, O.: Speech emotion recognition using attention model. Int. J. Environ. Res. Public Health **20**, 5140 (2023). https://doi.org/10.3390/ijerph20065140
9. Liu, G., Cai, S., Wang, C.: Speech emotion recognition based on emotion perception. EURASIP J. Audio Speech Music Process. **2023**, 22 (2023). https://doi.org/10.1186/s13636-023-00289-4
10. Khalil, R.A., Jones, E., Babar, M.I., Jan, T., Zafar, M.H., Alhussain, T.: Speech emotion recognition using deep learning techniques: a review. IEEE Access **7**, 117327–117341 (2019). https://doi.org/10.1109/ACCESS.2019.2936124
11. Madanian, S., et al.: Speech emotion recognition using machine learning — a systematic review. Intell. Syst. Appl. **20**, 200266 (2023). https://doi.org/10.1016/j.iswa.2023.200266
12. Leelavathi, R., Aruna Deepthi, S., Aruna, V.: Speech emotion recognition using LSTM. Int. Res. J. Eng. Technol. (IRJET) **09**(01) (2022)
13. Atila, O., Şengür, A.: Attention guided 3D CNN-LSTM model for accurate speech-based emotion recognition. Appl. Acoust. **182**, 108260 (2021). 0003–682X/@ 2021 Elsevier Ltd.
14. Zhao, J., Mao, X., Chena, L.: Speech emotion recognition using deep 1D & 2D CNN LSTM networks. Biomed. Signal Process. Control **47**, 312–323 (2019)
15. Aouani, H., Ayed, Y.B.: Speech emotion recognition with deep learning. Procedia Comput. Sci. **176**, 251–260 (2020). In: 24th International Conference on Knowledge-Based and Intelligent Information & Engineering Systems, SCI
16. Shahsavarani, S.: Speech emotion recognition using convolutional neural networks. Spring 3-16-2018, Theses, Dissertations, and Student Research
17. CREMA-D, SAVEE], Your Machine Learning and Data Science Community/Datasets (www.kaggle.com)

Real Time Voice Language Interaction

Gagan S. Yadav[✉], J. Vimala Devi, Tanmai Jain, H. D. Harshith, and B. N. Neha

CSE, Dayananda Sagar College of Engineering, Bangalore, India
gaganyadav2002@gmail.com, jvimaladevics@dayanandasagar.edu

Abstract. Communication across linguistic boundaries is critical in an era of increased global connectivity and diverse cross-cultural interactions. This paper discusses the difficulties presented by traditional language translation techniques, emphasizing the importance of real-time voice interaction to overcome delays and facilitate natural speech flow. Recognizing the complexities of adult language acquisition, we proposed a Speech-to-Speech Translation (S2ST) system that integrates advanced machine learning models seamlessly via a socket-based communication application. We have used the three key models: OpenAI Whisper for Speech-to-Text, Google Deep Translator for Text-to-Text translation, and Google Translator for Text-to-Speech translation. Together with a sophisticated socket-based application, these models form an innovative real-time voice language interaction platform. Users can join rooms, select their preferred language, and engage in multilingual conversations with speech that is seamlessly translated in real time. The proposed S2ST system aims to bridge linguistic gaps in a variety of fields, including business, tourism, education, and administration. Unlike traditional manual translation methods that are limited to specific content types, proposed approach addresses a wide range of communication requirements, fostering a more inclusive and accessible global dialogue.

Keywords: Voice-to-Voice translation · Real time communication · Sockets · Delay · Language Interaction

1 Introduction

In an era of increased global interaction, effective communication across languages is crucial. Conventional translation methods often cause delays, hindering natural speech flow. Learning a new language as an adult is found to be more challenging, making speech-to-speech technology valuable. While manual translation has been applied to specific content, there is a significant backlog for broader applications like business, tourism, and education.

Interest in Speech-to-Speech Translation (S2ST) has surged due to the growing demand for translingual communication. This technology facilitates access to multilingual information, easing communication among individuals speaking different languages. S2ST plays a vital role in breaking down linguistic barriers in international trade and cross-cultural interactions.

However, existing solutions relying on centralized servers introduce latency and privacy concerns. Real-time voice translation faces challenges in accuracy, especially in noisy environments. The noticeable lags in communication can affect the organic flow of conversations.

In the field of real-time speech-to-speech translation enabled by a socket-based communication application, we have used three essential models collaborate to produce a smooth and effective dialogue experience. The application's core models—the whisper model, deep translator, and google translator has helped improve overall accuracy, context awareness, and natural-sounding speech synthesis.

1.1 Speech-To-Text Model

A sizable language model for speech-to-text and translation is called OpenAI Whisper. It was trained using a sizable text and audio dataset and is based on the Transformer architecture. Whisper can translate audio into more than 100 languages and transcribing audio into more than 50 languages. Additionally, it can translate and transcribe noisy or multi-speaker audio.

1.2 Text-To-Text Model

Google AI created a machine translation model called Deep Translator. It was trained using a sizable text and code dataset and is based on the Transformer architecture. Text can be translated into more than 100 languages using Deep Translator. It is renowned for having excellent fluency and accuracy.

1.3 Text-To-Speech Model

Google created Google Translator, a web-based machine translation tool. It was trained using a sizable text and code dataset and is based on the Transformer architecture. More than 100 languages can be translated into text using Google Translator. Additionally, it can translate speech into more than 30 languages.

1.4 Application Using Sockets

The real-time application contains the combination of front-end using react and the back-end of flask with the usage of socket technology. The application creates a room for the users to join and choose their preferred language so as it receives the speech with that desired language.

2 Existing System

In [7], the authors address a crucial issue in S2ST, emphasizing the loss of prosody during translation. Their novel approach involves integrating source audio characteristics like pitch and energy into target language speech synthesis for improved prosody.

In [6], the paper explores challenges in language pairs, including linguistic differences and a lack of parallel data. To tackle these issues, the authors investigate various machine translation models, such as phrase-based and neural machine translation (NMT) approaches.

The author in [4] conducts a comparative study on using discrete speech units in speech recognition, translation, and comprehension, focusing on sub-word units like phonemes and graphemes.

The goal of [2] is real-time language translation using NLP with real-time audio input. The proposed system translates English audio into Indian language using OpenNMT and NMT.

Paper [10] compares WebSocket and TCP protocols, finding TCP outperforming WebSocket in terms of network traffic and data transfer time. WebSocket, despite a slight performance hit, is effective for long-running Web sessions requiring continuous data streaming between client and server.

Table 1. Literature Survey.

References	Objective of Proposed Work	Methodology	Accuracy Measures	Limitations
Transformer Based Direct Speech-To-Speech Translation with Transcoder [1]	Analysing and aligning lengthy speech sequences in both source and target languages, as well as generating extensive speech sequences from text	Improving the overall translation quality by optimizing the entire system instead of training and tuning separate components	Word Error Rate (WER) Character Error Rate (CER)	Syntactic differences impact model performance, with variations depending on source and target language similarities
Real Time Machine Translation System for English to Indian language [2]	The paper aims to present a Real Time Machine Translation System for English to Indian languages using NLP	The process involves preparing the input, breaking it into tokens, extracting features while filtering out noise, and then mapping sentences into the target language	Predicted average score, Perplexity	The main challenge for the proposed system will be to work on larger datasets in the future
LibriS2S: German English Speech-to Speech	Proposes Text to-Speech models that integrate source language information to	Enhance Fast Speech 2, a state-of-the-art TTS system, by integrating source text and	Mean Opinion Score (MOS) Pitch moments	The scarcity of parallel data with the same prosody or emotion in both languages is a

(*continued*)

Table 1. (*continued*)

References	Objective of Proposed Work	Methodology	Accuracy Measures	Limitations
Translation Corpus [7]	improve the quality of the generated speech	audio features to improve synthesized speech quality	Mean Absolute Error (MAE)	challenge in building TTS models that can use the audio from the source language to improve the generated speech in the target language
Large-Scale Streaming End-to-End Speech Translation with Neural Transducers [3]	Aims in reducing inference latency and improving accuracy compared to traditional cascaded ST model	Encoder translates input speech into hidden representations, the Prediction Network foresees the output text sequence, and the Joint Network combines information from the encoder and prediction network to generate the final translation output	Inference latency Word Error Rate (WER)	The proposed model may not be suitable for all types of speech translation tasks, and further research is needed to evaluate its performance on different datasets and in different settings
Listen attend spell [8]	A neural network that improves upon previous attempts to design models that are trained end-to-end for speech recognition, and to report on experiments conducted to evaluate the performance of the LAS model	Based on the sequence-to-sequence framework Listener is a pyramidal acoustic RNN encoder while the speller is a character-level RNN decoder. LAS model is trained end-toned using a combination of CTC and cross entropy loss functions	Word error rate (WER) Character Error Rate (CER)	Connectionist Temporal Classification (CTC) assumes label independence. Limiting transcript diversity sequence to-sequence models with attention have primarily been applied to phoneme sequences

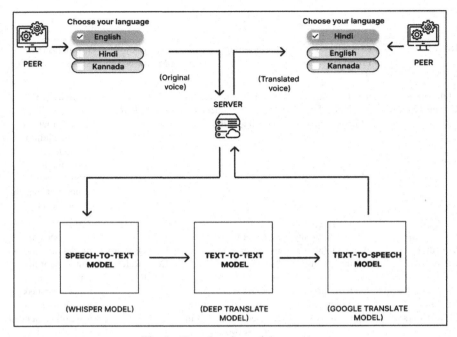

Fig. 1. Complete flow of the process

3 Methodology

3.1 Complete Flow of the Process

The process starts with the user entering a room and choosing their preferred language, which marks the first steps in the communication framework. Following that, the user engages in spoken communication in the language of choice. This spoken language is then seamlessly converted into text using a speech-to-text model, forming a critical stage in the communication pipeline. The converted text is then sent to a server, which is responsible for relaying this textual information to all peers in the shared space.

Upon receiving the text, the peers, who are spread throughout the room, use text-to speech models to translate and articulate the message in their native languages. Because each peer can comprehend and respond to the information in their preferred language, this multistep process ensures effective and inclusive communication among participants. The culmination of this intricate system occurs when peers speak aloud the translated text, allowing the user to hear the content in a language of their choice. Furthermore, the user retains the ability to end the call or switch languages dynamically, providing flexibility and user control within the communication platform. In essence, the flowchart (Fig. 1) depicts the entire sequence of events, from user entry to the final auditory reception of translated text via a collaborative and technologically mediated environment.

3.2 Speech-To-Text Conversion

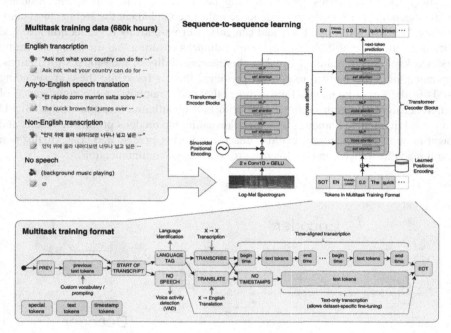

Fig. 2. The flow of Whisper model [12].

Whisper is an OpenAI general-purpose speech recognition model. It is a multitasking model that can recognize multilingual speech, translate speech, and identify languages. It was trained using a large amount of audio. Whisper is a traditional encoder-decoder transformer with 12 transformer blocks in both the encoder and the decoder. A self-attention layer and a feed-forward layer are included in each transformer block (Fig. 2).

A cross-attention layer is used to establish connectivity between the encoder and decoder. This layer enables the decoder to concentrate on the encoder output, assisting in the generation of text tokens that correspond to the audio signal. Whisper is trained on a massive dataset of multilingual audio and text data to improve its transcription capabilities. This extensive training data enables Whisper to transcribe speech proficiently in a variety of languages and accents, even in noisy environments [9].

3.3 Text-To-Text Conversion

A novel sequence-to-sequence neural machine translation (NMT) model developed with the Keras framework and the TensorFlow backend. This architecture uses embedding layers, LSTM encoders, Repeat Vector layers, and LSTM decoders with SoftMax activation to capture and translate complex language patterns (Fig. 3). The model's high accuracy in language translation across diverse benchmark tests is demonstrated by experimental results, positioning it as a versatile and scalable contribution to the field of machine translation.

The process emphasizes a robust training methodology, employing tokenization and padding techniques for uniform sequence lengths to complement the architectural components. The RMSprop optimizer and categorical cross-entropy loss function are used to train the model, with Model Checkpoint callbacks ensuring that the best-performing model is kept during training. Notably, the interpretability of the model's predictions, implementing a mechanism to convert predicted indices back into the corresponding words in the target language. Furthermore, the study emphasizes the significance of comprehensive evaluation metrics, considering both quantitative measures such as BLEU scores and qualitative human assessments, revealing the model's promising results and potential applicability in real-world scenarios requiring precise and contextually rich neural machine translation for effective cross-language communication.

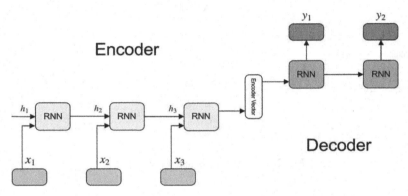

Fig. 3. Implementation of LSTM encoder-decoder [13].

3.4 Text-To-Speech Conversion

Google Translate is a multilingual neural machine translation service created by Google that allows users to translate text, documents, and websites from one language to another. It provides a website interface, an Android and iOS mobile app, and an API that allows developers to create browser extensions and software applications. Because its algorithms are based on statistical or pattern analysis rather than traditional rule-based analysis, Google Translate does not use grammatical rules.

Google Text-to-Speech (TTS) is a technology that converts text into spoken voice output. To generate high-quality voice output from text input, the technology employs a combination of machine learning algorithms and speech synthesis techniques [11]. Various techniques and algorithms that the system employs:

a. Text analysis is the first step in the Google TTS process:

i. Language Identification: Using a combination of natural language processing (NLP) techniques such as n-gram models, POS tagging, and named entity recognition, the system determines the language of the input text. ii. Dialect Identification: After identifying the language, the system determines the dialect of the input text. This is accomplished by examining the text's syntax, semantics, and phonology.

iii. Pronunciation Analysis: The system analyses the pronunciation of the input text while considering the language's phonetic and phonological rules. b. Voice Model Selection:

The system chooses the appropriate voice model for the text based on an analysis of the input text. Google TTS employs several voice models, each with distinct characteristics such as voice tone, pitch, and speed. The voice models are trained on large datasets of human speech, allowing them to learn human speech patterns and characteristics. The system chooses the voice model that best matches the input text's language, dialect, and pronunciation. c. Synthesis:

The system uses the voice model that was chosen to synthesize the text into voice output. The audio signal for the text is generated using a combination of signal processing techniques and machine learning algorithms in this process. Several sub-steps are involved in the synthesis process, including:

i. MFCCs (Mel-Frequency Cepstral Coefficients) Extraction: The system extracts the MFCCs from the input text, which represent the speech's spectral characteristics.
ii. Spectrogram Generation: The system generates a spectrogram from the MFCCs, which is a visual representation of the audio signal. iii. Vocoder Synthesis: The system generates the audio signal from the spectrogram using a vocoder, which is a type of digital synthesizer. iv. Post-processing: The synthesized audio signal is then enhanced in quality and clarity. This may entail adjusting the audio signal's volume, pitch, and speed, as well as applying various effects to improve its overall sound quality

The final output is then delivered to the user via their device's speakers or headphones. The system's machine learning algorithms and speech synthesis techniques produce high-quality voice output that sounds natural and authentic

3.5 Real-Time Application

This involves the creation of a multilingual web application that allows users in a shared virtual space to communicate in real time and translate languages. WebSocket technology is used in the application to establish persistent and bidirectional communication channels between the server (implemented with Flask) and clients (implemented with React). The ability for users to join virtual rooms, select their preferred language, and engage in conversations with peers speaking different languages is the application's central feature. WebSocket connections allow for the instant transmission of speech-to text converted messages, creating a dynamic and interactive platform for multilingual communication. The importance of incorporating WebSocket technology stems from its ability to support seamless, low-latency communication, fostering a collaborative environment in which users can interact in real time while overcoming language barriers within shared virtual spaces [10].

4 Results

4.1 Output of Speech-To-Text Model

Figure 4 shows the accuracy of speech-to-text conversion based on the Whisper model.

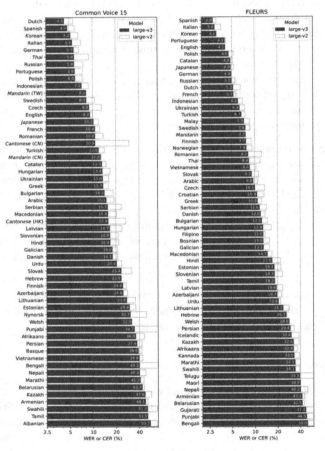

Fig. 4. Word Error Rate of Speech-to-text Model

4.2 Output of Text-To-Text Model

Figure 5 depict the accuracy of the Deep Translate model based on Bilingual Evaluation Understudy BLEU score which is a metric for assessing the quality of machine-generated text. The BLEU score compares machine-translated sentence n-grams to human translated sentence n-grams.

    a. Hindi-English conversion:
        i. Neural Translation model:
            Average BLEU Score: 0.222493077743151 ii.
        Deep translator:
            Average BLEU Score: 0.2757435129208912
    b. English-Hindi conversion:
        i. Neural Translation Model:
            Average BLEU Score: 0.4161874116489648 ii.
        Deep translator:
            Average BLEU Score: 0.2933072937797113

4.3 Output of Text-To-Speech Model

Figure 5 shows the accuracy against the words per rate of the Google translation model.

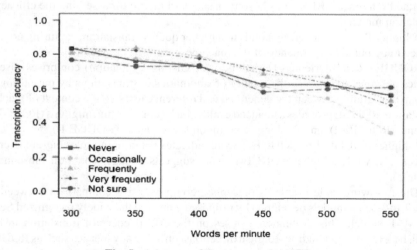

Fig. 5. Accuracy of Text-to-speech model.

4.4 The Real-Time Communication Application

- The home page allows user to first select his/her preferred language and then join in a particular room to start the conversion.

- Any number of users can enter the room and converse in their preferred language easily. One must click on the mic and start speaking.
- Each user hears translated text spoken by peers in the language of their choice with accuracy and a minimal delay. The user can also end the call and switch languages.

5 Table of Comparison

The following table (Table 2) presents a comprehensive comparison of text-to-text translation performance between the Deep Translator and Neural Machine Translation (NMT) models in the Real-Time Voice Language Interaction system. Evaluation metrics, including Average BLEU Score, ROUGE-1 F-score, ROUGE-2 F-score, and ROUGE-L F-score, are utilized to objectively assess the accuracy, fluency, and contextual relevance of translated text. The table specifically focuses on English-to-Hindi and Hindi-to-English translation directions, providing insights into the strengths and weaknesses of each model in handling these language pairs. These metrics play a pivotal role in quantifying the success of the Real-Time Voice Language Interaction system in breaking down language barriers and facilitating seamless cross-language communication (Figs. 6, 7, and 8).

Metrics used:

The BLEU (Bilingual Evaluation Understudy) score is a widely utilized metric in the evaluation of machine translation, offering a quantitative measure of the similarity between machine-generated translations and reference translations. Originally designed for evaluating the quality of machine-generated translations in comparison to human generated references, BLEU has become a standard metric for assessing the efficacy of translation models.

A higher BLEU score corresponds to a higher quality translation, capturing not only the accuracy but also the fluency of the machine-generated output.

ROUGE (Recall-Oriented Understudy for Gisting Evaluation) comprises a set of metrics designed to evaluate the quality of summaries or translations by measuring the overlap of n-grams between the generated and reference texts. In the context of machine translation, ROUGE provides a nuanced evaluation by considering unigrams (ROUGE1), bigrams (ROUGE-2), and the longest common subsequence (ROUGE-L).

A higher ROUGE-1 or ROUGE-2 score indicates better unigram or bigram overlap, respectively, while a higher ROUGE-L score suggests a longer common subsequence (Table 1).

The following table (Table 3) presents a comprehensive comparison between the speech-to-speech models developed by proposed team and the widely recognized Seamless M4T model. This evaluation focuses on the effectiveness of translation in both English to-Hindi and Hindi-to-English directions, utilizing key metrics such as ROUGE-1 F-score and Average BLEU. These metrics provide insights into the fluency, accuracy, and overall performance of the models, allowing for a nuanced assessment of their respective capabilities in facilitating seamless cross-language communication.

Table 2. Comparison of text-to-text translation performance between the Deep Translator and Neural Machine Translation (NMT) model.

Translation Direction	Model	Average BLEU Score	ROUGE-1 F-score	ROUGE-2 F-score	ROUGE-3 F-score
English to Hindi	Deep Translator	0.4034	0.6939	0.5311	0.6739
English to Hindi	NMT Model	0.4162	0.8137	0.6270	0.8047
Hindi to English	Deep Translator	0.4803	0.8711	0.6984	0.8511
Hindi to English	NMT Model	0.3629	0.7959	0.5790	0.7747

Table 3. Comparison of text-to-text translation performance between the Deep Translator and Neural Machine Translation (NMT) model.

Translation Direction	Metric	Seamless M4T	Proposed Model
English to Hindi	ROUGE-1 F-score	0.4301	0.5786
	Avg BLEU	0.1862	0.1595
Hindi to English	ROUGE-1 F-score	0.6126	0.6399
	Avg BLEU	0.1959	0.2890

Fig. 6. Results of English-to-Hindi translation of proposed model and Seamless M4T model.

	hindi	english	m4t	our_model
0	राजनीतिज्ञों के पास जो कार्य करना चाहिए, वह करने कि अनुमति नहीं है	politicians do not have permission to do what needs to be done.	Politicians are not allowed to do what they are supposed to do.	Politicians don't have permission to do what they should be doing.
1	मई आपको ऐसे ही एक बच्चे के बारे में बताना चाहूँगी.	I'd like to tell you about one such child,	I would like to tell you about such a child.	I would like to tell you about one such child,
2	यह प्रतिशत भारत में हिन्दुओं प्रतियत से अधिक है।	This percentage is even greater than the percentage in India.	This percentage is higher than the percentage of Hindus in India.	This percentage is more than the percentage of Hindus in India.
3	हम ये नहीं कहना चाहते कि वो ध्यान नहीं दे पाते	what we really mean is that they're bad at not paying attention.	We don't want to say that he can't pay attention.	We don't want to say that they can't pay attention
4	इन्हीं वेदों का अंतिम भाग उपनिषद कहलाता है।	The ending portion of these Vedas is called Upanishad.	The last part of these Vedas is called Upanishad.	The last portion of these Vedas is known as Upanishad.
5	कश्मीर के तत्कालीन गवर्नर ने इस हस्तांतरण का विरोध किया था, लेकिन अंग्रेजी की सहायता से उनकी आवाज़ दबा दी गयी	The then Governor of Kashmir resisted transfer, but was finally reduced to subjection with the aid of British.	The then governor of Kashmir had opposed the transfer, but with the help of the British, his voice was suppressed.	The then Governor of Kashmir had opposed this transfer, but with the help of the British his voice was suppressed.
6	इसमें तुमसे पूर्व गुज़रे हुए लोगों के हालात हैं।	In this lies the circumstances of people before you.	This is the condition of the people who passed before you.	It contains the circumstances of the people who came before you.
7	और हम होते कौन हैं यह कहने भी वाले कि वे गलत हैं	And who are we to say, even, that they are wrong	And who are we to even say that they are wrong?	And who are we to say that they are wrong
8	ग्लोबल वॉर्मिंग से आशय हाल ही के दशकों में हुई वार्मिंग और इसके निरंतर बने रहने के अनुमान और इसके अप्रत्यक्ष रूप से मानव पर पड़ने वाले प्रभाव से है।	"Global Warming" refer to warming caused in recent decades and probability of its continual presence and its indirect effect on human being.	Global warming is the result of decades of global warming and its indirect effects on human beings.	Global warming refers to the warming that has occurred in recent decades and is projected to continue and its indirect effects on humans.
9	हो सकता है कि आप चाहते हों कि आप का नज़रनीमेंटड ड्राइवन किसी समर्थन के ह विशेष स्कूल, या किसी स्वतंत्र स्कूल में जाए, इसमसके पास विशेष शैक्षणिक ज़रूरतों वाले बच्चों के प्रति सहूलियत हो	You may want your child to go to a school that is not run by the LEA - a non-maintained special school or an independent school that can meet your child's needs.	You may want your child to go to a special school without any support, or to an independent school with special educational needs.	You may want your child to attend a special school without any support, or an independent school which has facilities for children with special educational needs.

Fig. 7. Results of Hindi-to-English translation of proposed model and Seamless M4T model.

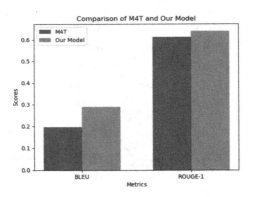

Fig. 8. Comparison of M4T and proposed model

6 Conclusion

This study presents a novel real-time voice language interaction system that integrates cutting-edge machine learning models within a socket-based application. The proposed method addresses the shortcomings of traditional translation methods, providing a flexible solution for effective communication across linguistic divides. The proposed system demonstrates improved accuracy, context awareness, and natural-sounding speech synthesis by combining the efforts of Speech-to-Text, Text-to-Text translation, and Text-to Speech models.

Proposed evaluation of the Real-Time Voice Language Interaction system focused on translation performance, emphasizing the Deep Translator and NMT models. For English to Hindi text translation, Deep Translator achieved a moderate BLEU score of 0.4034 and proficient ROUGE-1, ROUGE-2, and ROUGE-L F-scores (0.6939, 0.5311, and 0.6739). The NMT model outperformed with a slightly higher BLEU score (0.4162) and superior ROUGE scores (0.8137, 0.6270, and 0.8047), indicating improved fluency and accuracy.

In Hindi to English text translation, Deep Translator excelled with a BLEU score of 0.4803 and robust ROUGE scores (0.8711, 0.6984, and 0.8511). Despite a lower BLEU score (0.3629), the NMT model demonstrated competitive ROUGE scores (0.7959, 0.5790, and 0.7747), showcasing its proficiency in both translation directions.

In the comparative analysis between proposed speech-to-speech model and the Seamless M4T model, proposed system consistently demonstrates superior performance in both translation directions. For English to Hindi, proposed model achieves a higher ROUGE-1 F-score (0.5786) and, although exhibiting a lower Average BLEU (0.1595), showcases enhanced fluency and accuracy compared to Seamless M4T (ROUGE-1 F-score: 0.4301, Avg BLEU: 0.1862).

In the Hindi to English direction, proposed model outperforms Seamless M4T with a higher ROUGE-1 F-score (0.6399) and a substantially improved Average BLEU (0.2890) compared to the latter (ROUGE-1 F-score: 0.6126, Avg BLEU: 0.1959). These results emphasize the effectiveness of proposed speech-to-speech model in breaking down language barriers, fostering cross-language communication with improved accuracy, fluency, and contextual relevance.

In conclusion, the comparative evaluation of Deep Translator and NMT models underscores the importance of choosing the right model for specific translation needs. Proposed speech-to-speech model stands out prominently in the comparative analysis against the Seamless M4T model, consistently exhibiting superior performance in both English to Hindi and Hindi to English translation directions. Proposed Real-Time Voice Language Interaction system, enriched by these findings, provides a versatile platform for multilingual communication, and contributes to the larger goal of fostering global understanding and collaboration through seamless, real-time communication across languages. The pursuit of seamless cross-language communication remains an ongoing endeavour, and this research contributes valuable insights towards achieving that goal.

References

1. Kano, T., Sakti, S., Nakamura, S.: Transformer-based direct speech-to-speech translation with transcoder. IEEE Spoken Language Technology Workshop (SLT), pp. 958–965. Shenzhen, China (2021). https://doi.org/10.1109/SLT48900.2021.9383496
2. Vyas, R., Joshi, K., Sutar, H., Nagarhalli, T.P.: Real time machine translation system for english to indian language. In: 2020 6th International Conference on Advanced Computing and Communication Systems (ICACCS), pp. 838–842. Coimbatore, India (2020). https://doi.org/10.1109/ICACCS48705.2020.9074265
3. Jian, X., Wang, P., Li, J., Post, M., Gaur, Y.: Large-scale streaming endto-end speech translation with neural transducers. arXiv:2204.05352 [cs.CL] (2022)
4. Chang, X., Yan, B.: Speech recognition, translation, and understanding with discrete speech units: a comparative study. arXiv:2309.15800 (2023)
5. Gong, H., Dong, N.: Multilingual speech-to-speech translation into multiple target languages. arXiv:2307.08655 (2023)

6. Gandhi, V.A., Gandhi, V.B., Gala, D.V., Tawde, P.: A study of machine translation approaches for gujarati to english translation. In: 2021 Smart Technologies, Communication and Robotics (STCR), pp. 1–5. Sathyamangalam, India (2021). https://doi.org/10.1109/STCR51658.2021.9588859
7. Jeuris, P.: Jan Niehues LibriS2S: A German-English Speech-to-Speech Translation Corpus. arXiv:2204.10593 (2022)
8. Chan, W., Jaitly, N.: Listen, Attend and Spell. arXiv:1508.01211
9. Radford, A., Kim, J.W., Xu, T.: Robust speech recognition via large-scale weak supervision. arXiv:2212.04356 (2022)
10. Skvorc, D., Horvat, M., Srbljic, S.: Performance evaluation of Web socket protocol for implementation of full-duplex web streams. https://doi.org/10.1109/MIPRO.2014.6859715
11. Nwakanma, C., Oluigbo, I., Izunna, O.: Text – To – Speech Synthesis (TTS) (2014)
12. Radford, A., Kim, J.W., Xu, T., Brockman, G., McLeavey, C., Sutskever, I.: Robust Speech Recognition via Large-Scale Weak Supervision. arXiv:2212.04356
13. Understanding Encoder-Decoder Sequence to Sequence Model. https://towardsdatascience.com/understanding-encoder-decoder-sequence-to-sequence-model-679e04af4346

Exploring In-Context Learning: A Deep Dive into Model Size, Templates, and Few-Shot Learning for Text Classification

Areeg Fahad Rasheed[1](✉), Safa F. Abbas[2], and M. Zarkoosh[3]

[1] College of Information Engineering, Network Department, Al-Nahrain University, Baghdad, Iraq
`areeg.fahad@coie-nahrain.edu.iq`
[2] Department of Cybersecurity Engineering, College of Information Engineering, Al-Nahrain University, Jadriya, Baghdad, Iraq
`safa@alrasheedcol.edu.iq`
[3] Software Engineering, Baghdad, Iraq

Abstract. State-of-the-art large language models (LLMs) have undergone rigorous training on extensive datasets, resulting in the development of highly complex models with an extensive number of parameters. These advanced models have exhibited a remarkable ability to categorize and generate text tailored to specific tasks effectively, even when they have not been previously exposed to the pertinent data. This cutting-edge technology is commonly referred to as in-context learning (ICL). The primary objective of this research paper is to conduct an in-depth analysis of ICL, encompassing its applications, optimal methodologies, and the critical factors that impact its effectiveness, such as the template, model size, and shot types (i.e., the number of the example used). Our study involves a meticulous evaluation and implementation of the key ICL strategies across two distinct datasets, revealing the best practices for leveraging ICL to achieve high results and enhance overall performance.

Keywords: In context Learning · Natural Language Processing · Large Language Model · Deep learning.

1 Introduction

Artificial intelligence (AI) is one of the important technologies that will influence our future. It enables computers to perform tasks that used to need human intelligence, automating a wide range of industries, including manufacturing, entertainment, and healthcare as well as finance. Diverse technologies like machine learning, computer vision, and natural language processing collapse in the broadly defined field of artificial intelligence [18].

Large language models (LLMs) is one of the AI technology, LLMs have revolutionized numerous domains, showcasing exceptional performance in tasks such as text classification, code generation, and question-answering [6, 13]. However, the size and computational demands of high-performing LLMs often make them inaccessible to many users, requiring substantial resources and massive datasets for development and training [7]. Recently, various optimization methods have been introduced to reduce the computational resources required for training large language models. These methods include LoRA [5], distillation [11], quantization [3], pruning [12], and in-context learning (ICL) [4], which will be the primary focus of this paper.

2 In-Context Learning (ICL)

In-context learning (ICL) is a machine learning paradigm where a large language model (LLM) is able to learn a new task from a small number of examples provided at inference time [17, 24, 25]. This is in contrast to traditional machine learning approaches, where the model needs to be trained on a large dataset of labeled examples before it can be used to make predictions [15, 16, 19]. ICL has a number of advantages over traditional machine learning approaches. First, it allows LLMs to be used to solve new tasks without the need for fine-tuning. By leveraging this technique, we achieve drastic reductions in both resource usage and execution times. This makes LLMs more versatile and adaptable [9]. Second, ICL can be used to solve tasks where there is only a limited amount of labeled data available. This is important for many real-world applications, such as medical diagnosis and financial forecasting. Figure 1 illustrates the main components of ICL, which include the prompt, the context windows, the LLM model, and the completion. The prompt consists of three other components: the template, the examples, and the task required for classification [20].

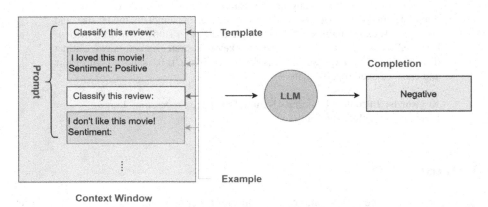

Fig. 1. In context learning the main components

In the following, a detailed explanation of each component is provided.

- Prompt: The prompt defines the overall goal and desired outcome of the ICL process. It specifies the task the LLM needs to perform, the format of the expected output, and any specific requirements or constraints [21]. This information guides the LLM towards generating a relevant and accurate response. The prompt consists of three components, as explained below:

- Template: The template serves as the opening statement that instructs the LLM about the required task [22]. This instruction varies depending on the specific task, such as classification, question answering, and others. For instance Classify the following sentence as positive or negative: {text}, Provide a concise summary of the following article: {text}, Translate the following sentence from English to Arabic: {text}
- Example (shot type): In-context learning heavily relies on examples within the prompt, often referred to as "shots," to guide the LLM toward the desired output. Three main shot types are commonly used, each with its own advantages and disadvantages.

  - Zero-shot learning: Zero-shot learning relies only on a prompt containing a template and the text of the required task, which may result in inaccurate or inconsistent outputs. Consequently, this approach may not be well-suited for complex tasks.
  - One-shot learning: The prompt in this case comes with multiple examples of the required task, providing a concrete illustration of the desired output and enhancing accuracy while reducing ambiguity. However, it may still lead to biased or inaccurate outputs if the provided example is not representative of the broader task.
  - Few-shot learning: The prompt includes multiple examples exceeding one. Provides the LLM with a broader range of reference points, leading to more accurate and consistent outputs. However, using multiple examples can exceed the context window's capacity and impact performance.

- Input (Required Task): This final element of the prompt specifies the text that needs to be processed and manipulated by the LLM.

- Context Windows: In large language models (LLMs), context windows refer to the maximum number of tokens (words) the model can process at once, including the prompt contents (templates, examples, and required tasks) [23]. These window sizes vary across different models. For example, GPT-3 has a context window of 2048 tokens, BARD has 1024, and BLOOM has 4096.
- The model acts as the ICL's central engine, consuming content within the context window. It learns from provided examples, analyzing and processing them to generate the desired completion, fulfilling the intended task.
- Completion: In the context of ICL, a completion is the final output generated by the model after it has processed the input and learned from the provided examples. The completion can take various forms, depending on the specific ICL and its intended function. It could be text, code, images, music, or any other format relevant to the task.

In this paper, we will leverage In-Context Learning (ICL) to comprehensively assess its performance in task classification. We will investigate the impact of model selection, prompt templates, and the number of in-context examples (zero-shot, one-shot, and few-shot) on ICL performance. We will also explore how these factors interact with each other. Our goal is to provide a comprehensive understanding of the strengths and limitations of ICL for task classification and to identify the best practices for using ICL in real-world applications. The paper is organized as follows: Sect. 2 explains the datasets and models used for the classification task. In Sect. 3, we describe the experimental setup, including the templates used, the types of ICL shots, and the datasets. In Sect. 4, we discuss the main results, and in the last section, we conclude.

3 Datasets and Models Overview

This paper investigates the performance of two versions of the FlanT5 model on two separate datasets. The details of the datasets and models are provided in the following section.

3.1 Datasets

To assess the efficacy of In-Context Learning (ICL) in task classification, we employed two distinct datasets characterized by varying structures, features, and sizes. The first dataset comprised 5,170 publicly available emails categorized into spam and non-spam [4, 10]. The second dataset consisted of freely accessible online content, distinguishing between sexist and normal categories it consists of 60000 samples [8]. Table 1 summarizes the key statistical details for both datasets, including the number of features (e.g., words) in the dataset vocabulary, the average number of words per sample (e.g., sentence), and the maximum number of words encountered in any single sample, which is crucial for configuring the context window size in subsequent analysis to ensure it does not exceed the maximum token limit.

Table 1. Data-sets proprieties

Property	Email Dataset	Offensive dataset
Number of all Samples	5170	60000
Number of test Samples	1034 1	12000
Number of samples in class A	3672(ham)	44670 (normal)
Number of samples in Class B	1498(spam)	15330 (sexist)
Number of unique features (words)	50615	55641
Average words per sample	228 1	24
Largest number of words in a sample	8862	58
The smallest number of words in a sample	1	1
The number of samples contains features that the average	1550	27693

3.2 FlanT5 Model

Flan-T5 is a factual language model from Google AI, trained on a massive dataset of text and code [2]. It is a fine-tuned version of the T5 model [1], a general-purpose text-to-text transformer model. Flan-T5 is specifically designed for factual tasks such as question answering, summarization, classification, and translation. Flan-T5 is available in four sizes: base with 250 M parameters, large with 780 M, XL with 3B parameters, and XXL with 11B parameters [14]. The main difference between the different sizes is the number of parameters that the model has. The more parameters a model has, the more complex it is and the more data it needs to be trained on. We have used the base and large versions of the Google FlanT5 model.

4 ICL System Overview

The workflow for In-Context Learning (ICL) is illustrated in Fig. 2. The initial step involves preparing the prompt, which, as mentioned in the previous section, comprises various components. The first component is the templates, and for each dataset, we utilized three different templates, as shown in Table 2. The second step involves selecting the shot type. ICL was evaluated with three shot types, representing the number of examples fed into the prompt. The shot type is associated with the context window's capacity. For one-shots, there is only one example that can be considered. However, for a few shots, there is no such limit, allowing us to use more examples, especially in tasks such as email spam classification, where more examples are used, and fewer in tasks of sexism classification, as examples in the sexist dataset contain more tokens than those in the email dataset. It's important to note that the examples in this work are exclusively drawn from the training dataset. After preparing the prompt with templates and examples, we add the required tasks, combining them before feeding them to the

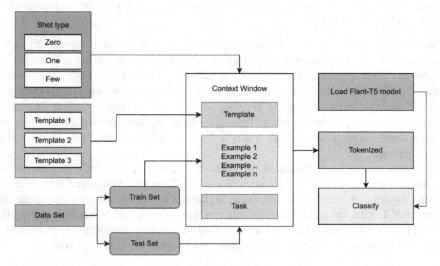

Fig. 2. ICL workflow

model. In this study, two different models were employed: FlanT5 Base and Large. The context window is tokenized and fed to the model for the classification process. The results are recorded for future analysis.

Table 2. Templates Used for Each Dataset

Template #	Email Dataset	Offensive dataset
Template 1	Email Classification: Message: {example} Label: {label}	Classify the following message as normal or sexist {example} label:{label}
Template 2	Classify the following mail as spam or human written: {example} label: {label}	Determine if this comment {example} contains sexist content or is normal: {label}
Template 3	classify the following mail spam or ham: {example} label: {label}	Analyze the following text to assess its potential for sexism or it is normal {example}:{label}

5 ICL Performance Evaluation and Discussion

We implemented two models, each tested with three different templates, and each template was evaluated with three shot types. In total, we conducted 18 experiments for each dataset. The performance metrics, including accuracy, F1 score, recall, and precision, were computed for our evaluation. Table 3 showcases the outcomes for spam mail detection, while Table 4 presents the results for sexist content detection. In the following, we provide a description of the effectiveness of each factor on the performance of In-Context Learning (ICL).

1. Model Size: As the results demonstrate, model size is the most significant factor influencing ICL performance. When using the smaller, base version of FlanT5, we achieved 75% accuracy on the email classification dataset and 56% on the sexist dataset. These results improved considerably with the larger FlanT5 model, with email spam detection accuracy jumping to 89% and sexism detection increasing to 77%. This improvement stems from the increased number of parameters in the larger model, which enhances its ability to learn complex patterns and achieve higher performance.
2. Template: As Table 3 illustrates, template choice can significantly impact performance. Template 1 led to a 74% accuracy, while Template 3 achieved 89% with the same model size. This stark difference highlights the power of template optimization in ICL. Carefully crafted templates can guide the system towards acquiring the most relevant features for a specific task. This makes them akin to hyper-parameters, requiring Careful fine-tuning and selection to optimize the learning process.

3. Shot type: The shot type refers to the number of examples used in the prompt. Increasing the number of examples typically improves system performance. This is because more examples enhance the language model's ability to understand the required task and adapt to its nuances. Conversely, using zero examples hinders this capability. When employing the model with zero-shot learning, we achieved 78\% accuracy for spam mail detection and 70% for sexist content classification. These figures improved when using few-shot learning: 89% for email and 77% for sexist content.
4. Context Windows: The context window, often referred to as the prompt text, consists of three components: the template, examples, and the text necessary for classification. The context window is typically limited to a specific word count, so allocating the available length effectively is crucial to accommodate all examples and text. This ensures that no essential information is omitted. Additionally, it is advisable to avoid using excessively long examples in the context model to prevent exceeding the word limit.

Table 3. Email Spam detection using FalnT5 base and large models results (%)

Temp #	Shot Type	FlanT5 base				FlanT5 large			
		Recall	Precision	F1-score	Accuracy	Recall	Precision	F1-score	Accuracy
Temp 1	Zero	70.00	64.59	58.64	70.62	70.72	79.30	58.69	70.72
	One	70.62	50.00	70.20	76.42	76.42	82.32	70.20	76.42
	Few	70.62	64.59	58.64	70.62	74.78	81.41	67.20	74.78
Temp 2	Zero	50.38	56.92	52.47	50.33	73.71	77.15	74.65	73.71
	One	60.28	70.38	61.99	60.28	79.51	79.13	77.48	79.51
	Few	74.49	77.29	75.31	74.49	84.83	84.45	84.45	84.83
Temp 3	Zero	54.87	67.72	56.50	54.87	78.93	78.11	78.25	78.93
	One	71.69	69.77	70.25	71.69	79.13	83.59	79.97	79.13
	Few	74.97	74.58	70.68	75.00	89.08	88.95	88.99	89.08

5. Not Constantly Optimal: In-context learning (ICL) can be a valuable choice when there is an insufficient amount of data available to train traditional models. However, ICL is not always the ideal solution, as there are cases where simpler models, such as machine learning models or other recurrent neural models, can achieve high performance, especially when the dataset is substantial. Additionally, when the length of the training sample exceeds the capacity of the context window, the system is constrained to process only the required task. In such instances, it becomes impractical to apply one or few shots. This limitation often arises when applying ICL to tasks involving long inputs like article classification or summaries.

Table 4. Sexist Text Classification using FlanT5 base and large models results (%)

Temp #	Shot Type	FlanT5 base				FlanT5 large			
		Recall	Precision	F1-score	Accuracy	Recall	Precision	F1-score	Accuracy
Temp 1	Zero	66.47	66.19	56.02	66.47	54.56	75.86	56.25	54.56
	One	67.97	67.24	57.36	67.97	59.56	76.38	61.69	59.56
	Few	70.59	66.17	54.31	70.59	76.60	74.82	75.23	76.60
Temp 2	Zero	64.85	60.40	48.00	64.85	57.82	75.99	59.87	57.82
	One	60.28	70.38	61.99	60.28	63.81	76.09	66.01	63.81
	Few	69.99	65.82	54.36	69.92	77.46	75.37	75.13	77.46
Temp 3	Zero	59.66	64.52	53.03	59.66	70.29	67.72	68.67	70.29
	One	66.55	66.13	55.93	66.55	75.38	72.67	72.90	75.38
	Few	70.18	66.97	56.15	70.01	76.87	75.20	75.59	76.87

6 Conclusion

This study investigated the performance of in-context learning (ICL) systems on two datasets: spam mail classification and sexist content identification. We examined four key factors influencing ICL: model size, templates, shot types, and context window size. Through 32 experiments, we evaluated the results using deep learning performance metrics like recall, precision, F1-score, and accuracy. Our findings revealed that model size had the most significant impact on ICL performance. Larger models consistently achieved higher accuracy. The structure of the prompt template also played a crucial role, followed by the number and variety of examples provided (shot types). Using multiple diverse examples within the prompt led to better performance than using fewer or no examples.

we suggest to investigating the ICL in diverse natural language processing tasks such as text generation, question answering, and summarization. Concurrently, we also propose exploring alternative models beyond the current utilization of the flat 5.

References

1. Adewumi, T., Sabry, S.S., Abid, N., Liwicki, F., Liwicki, M.: T5 for hate speech, augmented data, and ensemble. Sci **5**(4), 37 (2023)
2. Chung, H.W., et al.: Scaling instruction-finetuned language models. arXiv preprint arXiv:2210.11416 (2022)
3. Dettmers, T., Lewis, M., Belkada, Y., Zettlemoyer, L.: Llm. int8 (): 8-bit matrix multiplication for transformers at scale. arXiv preprint arXiv:2208.07339 (2022)
4. Mosbach, M., Pimentel, T., Ravfogel, S., Klakow, D., Elazar, Y.: Few-shot fine-tuning vs. in-context learning: A fair comparison and evaluation. arXiv preprint arXiv:2305.16938 (2023)
5. Ge, T., Hu, J., Wang, X., Chen, S.Q., Wei, F.: In-context autoencoder for context compression in a large language model. arXiv preprint arXiv:2307.06945 (2023)

6. Kaddour, J., Harris, J., Mozes, M., Bradley, H., Raileanu, R., McHardy, R.: Challenges and applications of large language models. arXiv preprint arXiv:2307.10169 (2023)
7. Kasneci, E., et al.: Chatgpt for good? on opportunities and challenges of large language models for education. Learn. Individ. Differen. **103**, 102274 (2023)
8. Kirk, H.R., Yin, W., Vidgen, B., Röttger, P.: Semeval-2023 task 10: explainable detection of online sexism. arXiv preprint arXiv:2303.04222 (2023)
9. Kojima, T., Gu, S.S., Reid, M., Matsuo, Y., Iwasawa, Y.: Large language models are zero-shot reasoners. Adv. Neural. Inf. Process. Syst. **35**, 22199–22213 (2022)
10. Kumar, N., Sonowal, S., et al.: Email spam detection using machine learning algorithms. In: 2020 Second International Conference on Inventive Research in Computing Applications (ICIRCA), pp. 108–113. IEEE (2020)
11. Li, L., Zhang, Y., Chen, L.: Prompt distillation for efficient LLM-based recommendation. In: Proceedings of the 32nd ACM International Conference on Information and Knowledge Management, pp. 1348–1357 (2023)
12. Ma, X., Fang, G., Wang, X.: LLM-pruner: on the structural pruning of large language models. arXiv preprint arXiv:2305.11627 (2023)
13. Min, B., et al.: Recent advances in natural language processing via large pre-trained language models: a survey. ACM Comput. Surv. **56**(2), 1–40 (2023)
14. Nicula, B., Dascalu, M., Arner, T., Balyan, R., McNamara, D.S.: Automated assessment of comprehension strategies from self-explanations using LLMS. Information **14**(10), 567 (2023)
15. Rasheed, A.F., Zarkoosh, M., Al-Azzawi, S.S.: The impact of feature selection on malware classification using chi-square and machine learning. In: 2023 9th International Conference on Computer and Communication Engineering (ICCCE), pp. 211–216. IEEE (2023)
16. Rasheed, A.F., Zarkoosh, M., Al-Azzawi, S.S.: Multi-cnn voting method for improved arbic handwritten digits classification. In: 2023 9th International Conference on Computer and Communication Engineering (ICCCE), pp. 205–210. IEEE (2023)
17. Xu, J., et al.: Unilog: Automatic logging via LLM and in-context learning. In: 2024 IEEE/ACM 46t International Conference on Software Engineering (ICSE), pp. 129–140. IEEE Computer Society (2023)
18. Hunt, E.B.: Artificial intelligence. Academic Press (2014)
19. Rasheed, A.F., Zarkoosh, M., Elia, F.: Enhancing graphical password authentication system with deep learning-based arabic digit recognition. Int. J. Inform. Technol. 1–9 (2023)
20. Bang, F.: Gptcache: an open-source semantic cache for LLM applications enabling faster answers and cost savings. In: Proceedings of the 3rd Workshop for Natural Language Processing Open Source Software (NLP-OSS 2023), pp. 212–218 (2023)
21. Zhou, Y., et al.: Large language models are human-level prompt engineers. arXiv preprint arXiv:2211.01910 (2022)
22. Arawjo, I., Swoopes, C., Vaithilingam, P., Wattenberg, M., Glassman, E.: Chain-forge: a visual toolkit for prompt engineering and LLM hypothesis testing. arXiv preprint arXiv:2309.09128 (2023)
23. Ratner, N., et al.: Parallel context windows for large language models. In: Proceedings of the 61st Annual Meeting of the Association for Computational Linguistics (Volume 1: Long Papers), pp. 6383–6402 (2023)
24. Rasheed, A.F., Zarkoosh, M., Abbas, S.F., Al-Azzawi, S.S.: Arabic offensive language classification: Leveraging transformer, LSTM, and SVM. In: 2023 IEEE International Conference on Machine Learning and Applied Network Technologies (ICMLANT), pp. 1–6. IEEE (2023)
25. Thirunavukarasu, A.J., Ting, D.S.J., Elangovan, K., Gutierrez, L., Tan, T.F., Ting, D.S.W.: Large language models in medicine. Nat. Med. **29**(8), 1930–1940 (2023)

Unlocking Complexity: Traversing Varied Hurdles in Punjabi Newspaper Recognition System

Atul Kumar[1]() and Gurpreet Singh Lehal[2]()

[1] Department of Computer Science, R.G.M. Govt. College Joginder Nagar, Mandi, Himachal Pradesh, India
atulkmr02@gmail.com

[2] Department of Computer Science, Punjabi University, Patiala, India

Abstract. The digitization of historical newspapers across diverse scripts encounters significant hurdles in achieving optimal OCR (Optical Character Recognition) accuracy. The efficacy of OCR systems is intricately linked to text segmentation, a process heavily influenced by the quality of the paper utilized in newspaper printing. This paper delves into the myriad challenges associated with the optical conversion of newspaper pages, with a specific focus on the distinctive issues encountered in the identification of Punjabi newspapers. Addressing these challenges is imperative for the accurate recognition of text in Punjabi newspapers. Several complications and unresolved problems unique to Punjabi newspaper identification are explored, emphasizing the necessity for tailored solutions. The paper also elucidates various techniques designed to mitigate these challenges, offering insights into potential resolutions. In this paper, we have compared 4 different newspapers of the Punjabi language to explore the difficulties in the Punjabi Recognition System. By navigating and resolving these obstacles, advancements in OCR accuracy for Punjabi newspapers can be achieved, facilitating the preservation and accessibility of historical information embedded in these valuable printed artifacts.

Keywords: Hurdles · Newspaper · OCR · Punjabi · Recognition

1 Introduction

Numerous organizations throughout the globe have started to digitize newspapers and make digital files accessible to the public with the rise of digital media. Newspapers preserve an extensive archive of the past. The best source of information is newspapers. According to studies from 2015, 90% of old newspapers were not digitalized. This evidence demonstrates that to digitize the newspapers, good work will be needed in this area. For Punjabi newspapers, the least amount of work has been put into digitizing them. Punjabi is the most spoken language in the region of Punjab and eastern Pakistan. Also, Punjabi is spoken in some parts of Canada. The official work in Punjab was done in the Punjabi language. Due to such high popularity of the language, most people in these regions read Punjabi newspapers. This heightened popularity underscores the importance of delving into the realm of text recognition for Punjabi newspapers and understanding the diverse challenges associated with this recognition process. [1] surveyed that the recognition accuracy rate ranges from 68% (character accuracy, no correction) for 350,000 pages of early 20th-century newspapers to 99.8% (word correctness, manually corrected) for 700,000 newspaper pages. Even though many newspapers are digitized various challenges are observed in recognition of these newspapers especially in the Punjabi language as the least work has been done in recognition of these newspapers. The newspaper recognition system involves preprocessing, Layout analysis, classification, and recognition as shown in Fig. 1.

It is essential to understand the layout of a newspaper page to correctly recognize the newspapers. The complex layout of a newspaper page presents significant difficulties in both page segmentation and text recognition. Scanned newspaper images may suffer from various degradations that reduce text recognition accuracy. While commercial OCR software is available for various document types, it often falls short when it comes to accurately segmenting and converting newspaper text, especially when newspaper images have degradations or other problems. For example, oversize newspaper formats pose challenges for conventional scanning equipment. Recognizing newspaper text is inherently more challenging than OCR for other documents, and this difficulty is compounded when dealing with historic newspapers due to language and typography variations. The rest paper is divided into various parts In Part 2, we discuss the Literature Review, In Part 3, we discuss challenges in the Recognition of Punjabi Newspapers, Part4 is Analysis and Discussion and we conclude in Part 5.

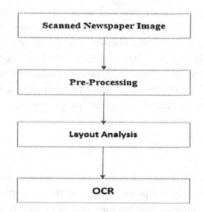

Fig. 1. Newspaper Recognition System workflow

2 Literature Survey

Newspapers contain a substantial amount of information that should be saved for later use. The themes that can be preserved from newspaper material are numerous and include data interpretation, examining past happenings, reviewing present legislation, political discussions, and many other things. [2] discussed various characteristics of the Gurumukhi script. They have created the linked list of characters in Punjabi language which are similar in shape. Different words were corrected in OCR recognition of the Punjabi Language increasing the accuracy up to 97.34%. [3] focused on creating corpora for old newspapers. Also, a pipeline-based approach to corpus creation that combines image processing, OCR, and filtering and recommended incorporating OCR mistake correction to raise the quality. For online Gurumukhi OCR, class labelling was done [4] on a word level. CNN architecture was used for training. As per analysis [5], the segmentation of characters in the Gurumukhi script is a very challenging task and different methods were discussed to accomplish this challenging task. [6] proposed a modification in AlexNet and GoogleNet and created a OCRNets for recognizing of urdu characters with the help of deep learning approach. They have used IFHCDB database datasets for experiments having 54 classes for Urdu. Tamil newspaper text was attempted to be divided up and recognized by [7]. Following the segmentation of lines, words, and characters using projection profiles, text, and graphic segmentation was carried out using RBF neural network and contour tracing. 157 segmented characters were identified using the RBF kernel. 94% accuracy was reported in this work. [8] proposed OCR for the Bangla language. The image was captured by using a flatbed scanner and various preprocessing was done on the image to clean the image. The classification was done based on the zonal information. An accuracy of around 95.50% was obtained at the word level. [9] proposed a system for recognizing newspapers in Bangla script using feature selection in conjunction with support vector machines. The accuracy of the overall system developed was 97.78. [10] proposed an innovative strategy utilizing two contemporary Recurrent Neural Network (RNN) models, namely Long-Short Term Memory (LSTM) and Bidirectional Long-Short Term Memory, to recognize online handwritten words in the Devanagari script [11]. Explored how diverse Optical Character Recognition (OCR) quality levels influence users' subjective perception. This investigation involved an interactive information retrieval task conducted on a digital collection of a historical Finnish newspaper that had undergone digital preservation. [12] used features to separate an image from the text. They used connected components to identify each object in the newspaper image and used various classifiers for training and testing purposes.

3 Methodology

In this part, we analyze the structure of Punjabi Newspapers. After that, we will discuss the features of the Gurumukhi script essential for the determination of challenges of recognition. After that, we interrogate various issues in the recognition.

3.1 Component of Punjabi Newspapers

A newspaper is typically structured into several key components as shown in Fig. 2, each serving a specific purpose in delivering information to readers. A newspaper typically comprises headlines introducing the article, often in larger font sizes, accompanied by subheadings. Digital pictures with captions illustrate the reported events. The main news content is presented in a block of text, with consistent font size within the body. News articles may vary in format, featuring single or multiple columns for headlines, text blocks, and images. Overall, these components work together to deliver information to readers in a structured manner.

Fig. 2. Various components in Punjabi Newspaper Image

3.2 Features of Gurumukhi Script

The term "Gurumukhi" originates from the expression "Gurumukhi," signifying "from the mouth of Guru." It pertains to the teachings and words conveyed by the Guru. The Gurumukhi script is employed for writing Punjabi, which serves as the official language in the Indian state of Punjab. It possesses distinctive features that define its character set.

- **Consonants (Akhar):** Gurumukhi script encompasses a total of 35 consonants, known as "Akhar." Among these consonants, the first three serve as the fundamental basis for vowels shown in Fig. 3.

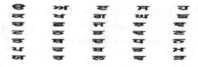

Fig. 3. Consonants in Punjabi language

- **Vowels (Laga Matras):** In addition to the consonants, the Gurumukhi script includes 10 vowels, referred to as "Laga Matras." These vowels function as modifiers, altering the pronunciation of consonants as shown in Fig. 4.

- **Vowel Modifiers:** The script also employs vowel modifiers, such as Tippi, Bindi and adhak indicate nasal sounds as shown in Fig. 5.

Sihari	Bihari	Kanna	Lavan	Dulavan
Dulaenkad	onkad	Hoda	knoda	Mukta

Fig. 4. Vowels in Punjabi language

Lagakhar	ਲਗਾਖਰ	Sign
Bindi	ਬਿੰਦੀ	ਂ
Tippee	ਟਿੱਪੀ	ੰ
Adhak	ਅਧਕ	ੱ

Fig. 5. Vowels modifier in the Punjabi language

ਸ਼, ਗ਼, ਖ਼, ਫ਼, ਲ਼, ਜ਼.

Fig. 6. Characters with dots in the Punjabi language

- **Characters with Dots:** Gurumukhi script includes six characters that are formed by adding a dot at the base of consonants. These characters are shown in Fig. 6.

- **Structural Consistency:** The writing style of the Gurumukhi script shares similarities with some other Indian scripts, particularly in terms of the "headline" feature. Headlines serve as the main horizontal lines above characters, connecting them to form complete words.

- **Character Structure:** Most Gurumukhi characters exhibit either full or half vertical sidebars, which is a defining structural feature. Additionally, these characters may contain open or closed loops, among various other distinguishing characteristics. OCR (Optical Character Recognition) processes involve multiple stages to accomplish their objectives. Before proceeding to the segmentation and recognition phases, it is essential to acquire a thorough understanding of the challenges encountered at each stage in achieving the desired outcome. The next section will cover all the challenges involved in recognition in Punjabi newspaper script

3.3 Various Problems in Recognition of Punjabi Newspapers

Enhancing the machine-readable text quality extracted from a newspaper page can be achieved by first identifying individual text blocks before the application of Optical Character Recognition (OCR). The quality of the paper on which the text is printed significantly affects the segmentation process. Newspapers printed in the Gurumukhi script often have lower paper quality compared to those in Roman scripts, adding complexity to segmentation, feature extraction, and classification. Several obstructions can occur during the segmentation and recognition of newspaper images. This obstruction results in many challenges. These include Complex newspaper layouts, touching character problems, mixing of scripts with numerals, Distorted character borders due to aging of paper, Poor printing quality, Marks on paper and color of paper, Shadowing on paper during scanning, Rigid paper binding, Coloured printing, and backgrounds, Mixing of maps, diagrams, and ligatures into newspaper pages.

Complex Newspaper Layouts. Nearly every newspaper article contains one or more images related to the event. Performing layout analysis of a document with multiple blocks is essential during the initial segmentation step. Images in a newspaper article can be located anywhere, and their sizes can vary according to the article's dimensions. Therefore, it is necessary to segment a newspaper article image into text blocks and non-text blocks before proceeding with OCR. Graphics, maps, and other elements may also be present within a newspaper article. Due to the complex layout of newspapers, layout analysis is a challenging task and involves a lot of effort.

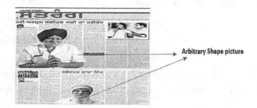

Fig. 7. Complex Layout Punjabi Newspaper Image

In Fig. 7, some of the pictures are wrapped within the text resulting in the images of complex layouts. The layouts are Manhattan which is rectangular in shape (Fig. 2) and non-Manhattan which may be any arbitrary shape (Fig. 7). The separation of text from non-Manhattan layouts is quite a challenging task. Techniques involved to solve this include Top-down approaches like RLSA [13], XY Cut [14], Bottom-up approaches Projection Profile-based methods and deep learning-based methods [15] like Deeplab V3 models [16], Faster RCNN [17], Mask RCNN [18].

Touching Character Problem. It is a very common problem when we talk about Gurumukhi script handling. A scanned document may contain pairs or longer sequences of characters that are touching in upper, lower, right, and left ends. This may occur because of too little space between neighboring characters or ink dispersion, due to the poor printing quality of the newspaper, where words and characters can come into close contact or even overlap, making it difficult to distinguish and segment them accurately. One common manifestation of this challenge is when the ink used in printing "bleeds," as shown in Fig. 8 which means that it spreads or smudges, causing characters and words to lose their distinct boundaries. This bleeding effect can result in blurred or distorted text, making it especially challenging for OCR algorithms to recognize and separate individual characters and words.

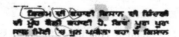

Fig. 8. Ink Bleeding causes touching characters

To address the challenge of touching words and characters in OCR for newspaper text, specialized techniques are required in the segmentation phase of the OCR process. These techniques aim to carefully separate words and characters that are in proximity or overlapping. This process may involve image analysis, edge detection, and contour-based approaches to identify and delineate the boundaries of each word and character. Different categories of characters that create touching problems are shown in Fig. 9.

(a) bindi (ò) touching with other characters.   ਜਿਵੇਂ ਜਵੇਂ ਅਸੀਂ

(b) adhak (ö) touching with other character   ਵਿਛ ਰੁੱਕ ਠੱਕ

(c) tippi (ö) touching with other character   ਸੰਖ ਪੰਖ ਸੰਜਮ

Fig. 9. Characters in newspapers touching with other character

ਸਿਨੇਲਸਰ ਸ਼ੀਲ ਦਾ ਪ੍ਰਸ਼ਸਰਤ ਤੇ ਰੁਮਾਂਚਿਕ ਸ਼ਹਿਰ ਜਲੰਧੀ ਪਾਸ ਤੋਂ ਵੀ ਸ਼ੁਰੂ ਹੁੰਦਾ ਹੈ। ਇਹ ਸ਼ੀਲ ਜਲੰਧੀ ਤੋਂ ਪੰਜ ਕਿਲੋਮੀਟਰ ਦੀ ਦੂਰੀ ਹੋਣ ਦੇ ਬਾਵਜੂਦ ਉਘਿਸ਼ਾਂਟ ਦਾ ਆਪਣਾ ਵਿਲੱਖਣ ਅਨੁਭਵ ਹੁੰਦਾ ਹੈ। ਇਹ ਲੇਖ ਉਥੇ ਤਕ ਪੁੱਜਣ ਦੇ ਅੰਹਿਸਾਸ ਨੂੰ ਬਿਆਨਦਾ ਅਤੇ

Fig. 10. Touching characters

Figure 10 shows how the characters of the lower line touch the characters of the upper line resulting in touching characters in red, which are very difficult to segment. Techniques based on Horizontal and vertical profile-based methods and detection of clusters are used to solve this problem.

Different Regions with Different Font Sizes. Multiple font sizes are frequently used in newspaper stories. For instance, the body text's font size is often higher than the headline sections, although image captions may have a smaller font size. Figure 11, where the body text font size is different, the headline font size is different, demonstrates these kinds of variations. This kind of variation in font size results in classification tasks difficult in OCR recognition. Techniques like binarization, RLSA, and Distance calculation between different lines of text are used to resolve this issue.

Fig. 11. Example of multiple fonts and size

Distribution of Headlines Across Multiple Columns. The placement of headlines across many columns is one of the significant difficulties faced in newspaper OCR systems. This situation frequently occurs in newspaper images with intricate design. Headlines, which frequently act as the first sentences of news items, frequently take up numerous columns in the body of the piece. In some instances, the headline text separates itself from the regular body text by spanning the breadth of many columns horizontally. Newspaper headlines frequently include many columns, which adds to the newspaper's overall presentation and style. To enable effective processing, managing text segmentation and recognition in newspaper stories with multi-column headers requires handling. Failure to address this challenge can result in the misalignment of segmented text, affecting the readability and overall quality of OCR output. Efficiently detecting and managing headlines that span multiple columns is a critical aspect of developing robust OCR solutions for newspapers with complex layouts. This challenge underscores the need for advanced layout analysis and segmentation techniques to accurately identify and process headlines within such articles. Figure 12 shows the spanning of headlines over multiple columns making it challenging to segment newspaper images. This can be overcome by segmenting headlines and news bodies from each other. Algorithms like RLSA, XY Cut, and Deep learning techniques are used to solve this issue.

Fig. 12. Headlines distributed over multiple columns

Spacing Between Words. Within a text line, the spacing between words may exhibit variations, with some word gaps appearing significantly larger than others as shown in Fig. 13. This discrepancy in word spacing is often intentional, introduced to justify the text within a line or to align it with a specific formatting requirement. Due to the wide range of word gaps, typical column segmentation techniques that simply rely on inter-word gaps may yield unreliable findings. If the algorithm interprets these extended word gaps as column separators, it can partition the newspaper text into many columns when none are there.

To overcome this problem, advanced column segmentation algorithms that consider context and layout clues in addition to inter-word gaps must be created. OCR systems can more accurately identify between legal column borders and variations in word spacing intended for text justification by including other characteristics, such as the presence of structural elements like headers, graphics, or captions.

Fig. 13. Different spacing of words

Quality of Newspapers. Poor paper quality can refer to several problems, such as stains on the paper, the general decline of paper quality, and aging-related page yellowing as shown in Fig. 14. When the image is subjected to binarization—a crucial stage in OCR—these problems might cause a significant decrease in the visibility and legibility of the text. The degraded images can be improved by applying several methods in the state of the art [19–21].

Fig. 14. (a) Yellowish page (b) Degraded paper quality

Quality of Printing. The recognition task is influenced by the print quality as well. In cases of poor printing quality, characters may become illegible or distorted, posing a challenge for OCR systems in accurately recognizing the text. Characters may appear smudged, broken, or irregular, deviating from standard fonts. Ink bleeding occurs when the ink used in printing spreads or smudges, causing characters and words to lose their sharp edges and merge together as shown in Fig. 15. The characters are broken from some sides and thinner.

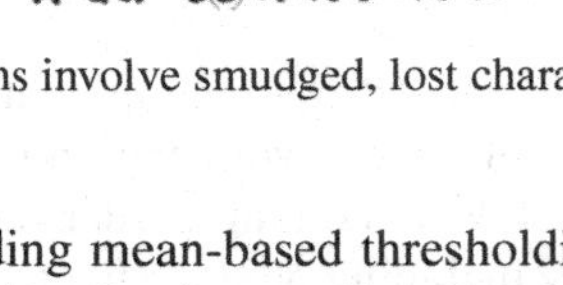

ਉਨ੍ਹਾਂ ਦੇ ਨਾਲ ਬੈਂਕ ਦੇ ਪ੍ਰਮੁੱਖ
ਸਿੰਘ ਸਮਰਾ ਸਾਬਕਾ ਕੈਬਨਿਟ
ਸਾਬਕਾ ਟਰਾਂਸਪੋਰਟ ਮੰਤਰੀ ।

Fig. 15. Marked sections involve smudged, lost characters, broken characters

Several techniques, including mean-based thresholding, chain codes, vertical projection profiles, and horizontal projection profiles, Deep neural forward networks [22] are used to detect or recognize the broken character in the image.

Skewing of Images. When working with scanned materials, such as newspaper stories, OCR (Optical Character Recognition) frequently encounters problems with skewed text in Fig. 16. Skew describes the non-horizontal slanting or tilting of text at an angle. This skew can be caused by several things, including uneven printing during the creation of the newspaper or incorrect document alignment during scanning. To guarantee accurate character identification in OCR, it is essential to correct slanted text. Skewed text can be caused by several things, including printing mistakes, paper alignment issues during printing, and scanning mistakes. These elements may cause the text to have an angular distortion. Because OCR algorithms normally presume that text is aligned horizontally, skewed text presents considerable difficulty to them. Characters may not be correctly detected by OCR if the text is still slanted, which could lead to erroneous results. Once the skew angle is determined, corrective measures are implemented to rotate the text and achieve horizontal alignment through skew correction procedures. Geometric transformations, such as rotation and shearing, are employed to correct skew in scanned images by applying mathematical processes to adjust text orientation. The Hough transform[23] effectively identifies lines, including text lines, allowing for the detection and correction of skew by analyzing dominating lines and their angles. Methods utilizing Fourier transform [24] convert image data from the spatial to the frequency domain, offering another approach for skew correction in image processing.

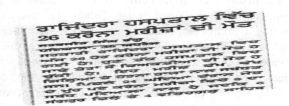

Fig. 16. Skewed Text in news article

Mixing of Numerals with the Script. In the case of Gurumukhi newspapers, there is news in which characters are mixed with some numeral shown in Fig. 17. Gurumukhi newspapers frequently include numerals to represent various types of information, such as dates, quantities, or statistics. These numerals are an integral part of the content and can appear within sentences or paragraphs. As a result, OCR systems need to distinguish between script characters and numerals accurately. The recognition of numerals must be context-aware, as the same character can represent a different value depending on its position within the text. Training over different kinds of characters other than the Gurumukhi script can solve the issue.

Distorted Characters. Sometimes in newspapers due to scanning and poor quality of paper, some of the characters are distorted. Ink spread over different parts is one of the major problems that arose due to the poor quality of newspaper papers as shown in Fig. 18. As a result, recognition of these characters becomes a challenging task.

Character edges are enlarged or contracted (dilated) or retracted (eroded) about a common background. An erosion technique can be used to make up for heavy ink

Fig. 17. Gurumukhi newspaper Image containing numerals

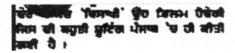

Fig. 18. Distorted Bleeding characters

bleeding from old papers. Characters can be reduced by erosion to their original glyph size.

Different Types of Fonts/Styles Used by Different Newspapers. A variety of Punjabi newspapers are available: Ajit Newspaper, Punjabi Tribune, Jagbani, and Punjabi Jagran as shown in Figs. 19a,b, and c, etc. All these newspapers have their unique layout, different font styles, and different printing paper and styles. Due to different styles, and layouts, printing paper styles to recognize text from these newspapers becomes very challenging and requires different techniques. To deal with this kind of issue is a very challenging task in recognition of Punjabi. Training over different fonts resolves this issue.

Fig. 19. Different newspapers having different styles, fonts (a)(b)(c)

4 Analysis and Discussions

In the previous section, we have presented different challenges that occurred during the recognition of Punjabi newspapers as shown in Table 1. Punjabi newspaper recognition is a very challenging task. These challenges encompass various aspects of OCR and the examination of documents specific to Punjabi script and newspapers. Addressing these challenges is crucial for accurately extracting machine-readable text from Punjabi newspaper images and improving the quality of digitized content. Table 1 shows how the different challenges of the Punjabi Newspapers can be resolved by using various techniques available in state of art literature. Various techniques are available in state of art literature to resolve the challenges and make a better recognition system.

Table 1. Techniques to resolve the problems of newspaper recognition system

S.No.	Problem	Techniques used to resolve the problem
1	Complex Layout	RLSA, Vertical and Horizontal Projection Profile, Deep learning Techniques, Connected component analysis
2	Touching character problem	Vertical and horizontal profiles, Contour and edge detection
3	Different Regions with Different Font Sizes	Training with different size of characters using Convolutional Networks
4	Quality of newspapers	Binarization, Image enhancement techniques
5	Skewing of images	Hough transform, Fourier transform, Geometric Transformation
6	Numerals mixing	Recognition techniques for different numerals
7	Distorted Characters	Dilation, Erosion
8	Different types of Fonts/Styles used by different Newspapers:	Training with different fonts

Table 2. Comparative Analysis of 4 Punjabi Newspapers for Recognition Difficulty

Abstract Newspaper	Script	Font Diversity	Punctuation	Availability of Annotated Data	Sample Size	Estimated Recognition Difficulty	Layout	Printing Quality
Punjabi Tribune	Gurmukhi	High	Limited	Low	Large	High	Complex	Good
Rozana Spokesman	Gurmukhi	High	Limited	Low	Small	Very High	Complex	Average
Punjabi Jagran	Gurmukhi	High	Limited	Low	Medium	Very High	Complex	Good
Ajit	Gurmukhi	High	Limited	Low	Small	Very High	Complex	Average

Table 2 compares four Punjabi newspapers (Punjabi Tribune, Rozana Spokesman, Punjabi Jagran, and Ajit) concerning their potential difficulty in automated recognition. The table considers the following factors:

- Script: All four newspapers use the Gurmukhi script, which has unique character shapes and presents distinct challenges for recognition systems compared to Latin-based scripts.
- Font Diversity: All newspapers utilize diverse fonts and styles, making it challenging for recognition models to generalize to unseen data. This is particularly relevant for newspapers like Ajit, known for its varied font usage.
- Punctuation: Traditional Gurmukhi scripts often lack standard punctuation, making it difficult to identify word boundaries and sentence structures. This poses a significant challenge for accurate text recognition and analysis.
- Availability of Annotated Data: Publicly available annotated datasets of Punjabi text are scarce, making it difficult to train robust recognition models. This lack of data hinders the development and performance of accurate recognition systems for all four newspapers.
- Sample Size: Newspaper availability can vary, with some like Punjabi Tribune having larger archives than others like Rozana Spokesman and Ajit. This limited sample size can further restrict the training of recognition models for specific newspapers.
- Estimated Recognition Difficulty: Based on the above factors, the estimated recognition difficulty is highest for Rozana Spokesman and Ajit due to their limited sample size and potentially higher font diversity. Punjabi Tribune and Punjabi Jagran face challenges due to the general difficulties of Gurmukhi recognition and limited annotated data, but their larger sample size might offer some advantages.
- Print Quality: All four newspapers have mixed fonts making it very difficult to recognize OCR. We have analyzed the print quality of four newspapers and found Punjabi Tribune and Punjabi Jagran have good print quality compared to others giving them an advantage in recognition.

All four newspapers present challenges for automated recognition due to the inherent complexities of the Gurmukhi script and the limitations of available resources. However, the degree of difficulty can vary depending on specific factors like font usage, data availability, and sample size.

5 Conclusion

In conclusion, the digitization and recognition of Punjabi newspapers pose a series of complex challenges due to the unique characteristics of the Gurmukhi script, the complex layouts of newspaper pages, and various other factors. In this paper we have discussed in detail challenges including complex newspaper layouts, touching character problems, variations in font sizes, distribution of headlines across multiple columns, differences in spacing between words, poor paper quality, characters and words touching each other, issues related to the quality of printing, skewed images, mixing of numerals with the script, distorted characters, and the use of different fonts and styles by different newspapers. We have also discussed future aspects by taking into account the current challenges in recognition. We have done a case study on four different Punjabi newspapers based upon various factors discussed in this paper to find the recognition difficulty in these newspapers. Due to widespread factors each newspaper has its features and challenges in recognition. This research will be very useful for doing research in the field of Newspaper OCR.

References

1. Sharma, H., Sethi, G.K. :Advancements in optical character recognition for Gurmukhi script. In :2019 Fifth International Conference on Image Information Processing (ICIIP), pp. 52–57. Shimla, India (2019)
2. Lehal, G.S., Singh, C., Lehal, R.: A shape-based post processor for Gurmukhi OCR. In: Proceedings of Sixth International Conference on Document Analysis and Recognition, pp. 1105–1109. Seattle, WA, USA (2001)
3. Klijn, E., Bibliotheek, K., Bonaparte, N.: The current state-of-art in newspaper digitization, D-Lib (2000)
4. Singh, S., Sharma, A., Chauhan, V.K.: Online handwritten Gurmukhi word recognition using fine-tuned Deep Convolutional Neural Network on offline features. Mach. Learn. Appl. **5** (2021)
5. Kumar, M., Sharma, R.K., Jindal, M. K., Singh, S.R.: Benchmark datasets for offline handwritten gurmukhi script recognition. In: Sundaram, S., Harit, G. (eds.) Document Analysis and Recognition. DAR 2018. Communications in Computer and Information Science, vol. 1020. Springer, Singapore (2018). https://doi.org/10.1007/978-981-13-9361-7_13
6. Aarif, M..K.O, Poruran, S.: OCR-Nets: Variants of pre-trained CNN for urdu handwritten character recognition via transfer learning. Procedia Comput. Sci. **171**, 2294–2301 (2020)
7. Aparna, K.H., Jaganathan, S., Krishnan, P., Chakravarthy, V.S.: An Optical character recognition system for Tamil newsprint. In: Proceedings of International Conference on Universal Knowledge and Language. pp. 881–886(2002)
8. Chaudhuri, B.B., Pal, U.: A complete printed bangla OCR system. Pattern Recogn. **31**, 531–549 (1998)
9. Ghosh, R.: Newspaper text recognition in Bengali script using support vector machine. Multimed. Tools Appl. (2023)
10. Ghosh, R., Keshri, P., Kumar, P.: RNN based online handwritten word recognition in devanagari script. In: 16th International Conference on Frontiers in Handwriting Recognition, pp. 517–522. Niagara Falls. USA (2018)
11. Kettunen, K., Keskustalo, H., Kumpulainen, S., Pääkkönen, T., Rautiainen, J.: Optical character recognition quality affects subjective user perception of historical newspaper clippings. J. Document. **79**(7), 137–156 (2023)
12. Jana, S., Das, N., Sarkar, R., Nasipuri, M.: Recognition system to separate text graphics from indian newspaper. In: Kar, S., Maulik, U., Li, X. (eds.) Operations Research and Optimization. FOTA 2016. Springer Proceedings in Mathematics and Statistics, vol. 225. Springer, Singapore (2018). https://doi.org/10.1007/978-981-10-7814-9_14
13. Stefano, F., Fabio, L., Fulvio, R., Floriana, E.: A Run Length Smooting-Based Algorithm For Non-Manhattan Document Segmentation (2012)
14. Sutheebanjard, P., Premchaiswadi, W.: A modified recursive x-y cut algorithm for solving block ordering problems. In: 2010 2nd International Conference on Computer Engineering and Technology, pp. V3-307–V3-311. Chengdu, China (2018)
15. Binmakhashen, G.M., Mahmoud, S.A.: Document layout analysis: a comprehensive survey. ACM Comput. Surv. **52**(6), 1–36 (2019)
16. Markewich, L., et al.: Segmentation for document layout analysis: not dead yet. IJDAR **25**, 67–77 (2022)
17. Kumar, A., Lehal. G.S.: A hybrid approach for complex layout detection of newspapers in gurumukhi script using deep learning. Int. J. Exper. Res. Rev. **35**, 32–42 (2023)
18. Long, S. Qin, S., Panteleev, D., Bissacco, A., Fujii, Y., Raptis, M.: Towards end-to- end unified scene text detection and layout analysis. In: 2022 IEEE/CVF Conference on Computer Vision and Pattern Recognition (CVPR), pp. 1039–1049. New Orleans, LA, USA (2022)

19. Shi, Z., Setlur, S., Govindaraju, V.: Image enhancement for degraded binary document images. In: 2011 International Conference on Document Analysis and Recognition, pp. 895–899. Beijing, China (2011)
20. Kumar, V., Bansal, A., Tulsiyan, G.H., Mishra, A., Namboodiri, A., Jawahar, C.V.: Sparse document image coding for restoration. In: 12th IEEE International Conference on Document Analysis and Recognition (ICDAR), pp. 713–717 (2013)
21. Pandey, R.K., Ramakrishnan, A.G.: Improving the perceptual quality of document images using deep neural network. In: Lu, H., Tang, H., Wang, Z. (eds.) Advances in Neural Networks. LNCS, vol. 11555. Springer, Cham.https://doi.org/10.1007/978-3-030-22808-8_44
22. Yetirajam, M., Ranjan, M., Chattopadhyay, S.: Recognition and classification of broken characters using feed forward neural network to enhance an OCR solution. Comput. Sci. (2012)
23. Hemantha, K.G., Shivakumara, P.: Skew detection technique for binary document images based on though transform. Int. J. Inf. Technol. **1**, 2401—2407 (2007)
24. Boiangiu, C.-A., Dinu, O.-A., Popescu, C., Constantin, N., Petrescu, C.: Voting-based document image skew detection. Appl. Sci. **10**, 2236 (2020)

Internet of Things (IoT)

Journal of Those ???

Patient Biomedical Monitoring - Remote Care System

D. Sri Chandana Charudatta[✉], B. Sidharth Reddy, CH. Pavan Sai, and S. Srinivas

Department of Electronics and Communication Engineering, Vardhaman College of Engineering, Hyderabad, Telangana, India
charudattasrichandana@gmail.com

Abstract. The Patient Biomedical Monitoring - Remote Care System is a project that aims to remotely monitor vital signs of patient. The system utilizes a remote monitoring device to collect data such as the patient's Electrocardiogram (ECG), Electroencephalogram (EEG), Electromyogram (EMG), Electrooculogram (EOG), body temperature, and Oxygen saturation (SPO2). This data is then analyzed and stored in an Excel sheet along with the patient's personal details and can be accessed by the monitoring system when needed. The system provides continuous monitor of the patient's vital signs remotely that aid the healthcare professionals. It allows early detection of any potential health issues and provide prompt health care needed by the patient. Additionally, the system can also give emergency alert to healthcare professionals for sudden anomalies in the patient's vital signs, allowing for swift medical attention. The Patient Biomedical Monitoring - Remote Care System has the potential to improve patient's status of health and speedy recovery by providing more efficient and effective healthcare.

Keywords: ECG · EEG · EOG · EMG · Remote monitoring system

1 Introduction

The use of electronic devices and IoT technology has significantly impacted the healthcare industry, revolutionizing the way medical professionals deliver care. These innovations have opened up new possibilities for remote monitoring, data collection, and real-time communication between patients and healthcare providers. With the advent of on-body devices, [10] such as fitness trackers, stress tracker and smartwatches, patients can now track their activity levels, vital signs and sleep pattern on a daily basis. These devices utilize sensors to capture data and transmit it wirelessly to healthcare systems, allowing healthcare providers to monitor patients' health remotely. This remote monitoring capability has proven especially valuable in managing chronic conditions like hypertension, diabetes, and heart disease. Furthermore, [4–6] IoT devices have been instrumental in improving patient safety and reducing medical errors. Smart medication dispensers can give an alert to patients so that they can take their medications at the correct times and can even send alerts to healthcare providers if doses are missed. In addition,

[7–9] IoT enabled medical equipment, such as smart infusion pumps and continuous glucose monitors, have built-in safety features and real-time monitoring capabilities that enhance patient care and outcomes. The integration of IoT devices in healthcare settings also facilitates the seamless exchange of information between different stakeholders, including patients, doctors, nurses, and caregivers. Electronic health records (EHRs) enable secure storage and sharing of patient data, promoting better coordination of care and reducing the risk of errors associated with manual record-keeping.

1.1 Role of Patient Biomedical Monitoring - Remote Care System in Addressing Healthcare Challenges

Continuous Monitoring of ICU Patients using ECG, EEG, EMG, EOG, Body Temperature, and SPO2 Sensor. In intensive care units (ICUs), the Patient Biomedical Monitoring - Remote Care System plays a vital role in continuously monitoring patients' health using a range of sensors. These sensors, including ECG, EEG, EMG, EOG, body temperature, and SPO2 sensors, provide real-time data on critical indicators, enabling healthcare providers to closely monitor patients' conditions. By utilizing ECG sensors, the system can capture and analyzed the electrical activity of the heart, monitoring for any abnormal rhythms or cardiac events. EEG sensors measure brain wave patterns, helping detect seizures or abnormal brain activity. EMG sensors assess muscle activity and can aid in the diagnosis and treatment of neuromuscular disorders. EOG sensors monitor eye movements and can be valuable in evaluating sleep patterns and detecting sleep disorders. Furthermore, the system incorporates body temperature sensors to track fluctuations in patients' temperature, which can be indicative of infections or other health concerns. SPO2 sensors measure oxygen saturation levels, enabling healthcare providers to assess respiratory function and detect any potential oxygenation issues. The continuous monitoring provided by the Patient Biomedical Monitoring - Remote Care System ensures prompt identification of any changes or abnormalities in these vital signs. This early detection allows healthcare professionals to intervene quickly, providing timely and appropriate care to ICU patients.

2 Literature Survey

2.1 Biomedical Methods for the Assessment of Health Parameters

This main section focuses on various biomedical monitoring techniques used to assess specific health parameters. These techniques include non-invasive tests for measuring oxygen saturation levels in the blood, a biomedical procedure for assessing muscle health and motor neurons, and a biomedical system for remote monitoring of heart rate, pulse rate, and respiration rate [1] introduces a biomedical procedure used to assess the health of muscles and the motor neurons that control them. This procedure is a valuable diagnostic tool in evaluating muscle-related conditions and monitoring patients' muscle health. By assessing muscle functionality and the communication between motor neurons and muscles, healthcare professionals gain critical insights into neuromuscular disorders, such as muscular dystrophy, myasthenia gravis, or peripheral neuropathies.

The procedure typically involves electromyography (EMG), a technique that records the electrical activity produced by muscles during contraction and at rest. EMG helps identify abnormal muscle activity, patterns, and potential issues with motor neuron function. By analyzing EMG signals, healthcare providers can determine the severity of muscle dysfunction, track disease progression, and develop targeted treatment plans tailored to the specific condition. This biomedical procedure serves as a valuable tool in assessing muscle health, diagnosing neuromuscular disorders, and monitoring the effectiveness of therapeutic interventions. In [2], a biomedical system is described, which incorporates ECG, PPG, and piezo sensors to measure heart rate, pulse rate, respiration rate, and transmit the data over a network for remote monitoring. This system revolutionizes healthcare by enabling real-time monitoring and remote transmission of vital signs data. The ECG sensor records the electrical activity of the heart, providing valuable insights into heart rate, rhythm, and abnormalities. The PPG sensor measures changes in blood volume in peripheral blood vessels, allowing for the calculation of pulse rate. Additionally, the piezo sensor detects chest wall movements during respiration, enabling accurate measurement of respiration rate. By integrating these sensors and transmitting the data over a network, healthcare providers can remotely monitor patients' vital signs in real-time. This capability is especially valuable for patients in home care settings, post-operative monitoring, or individuals with chronic conditions. Timely detection of any deviations or abnormalities in vital signs allows healthcare professionals to provide prompt interventions, adjust treatment plans, or initiate emergency measures when necessary. The biomedical system described in reference [2] exemplifies the potential of remote monitoring to improve patient outcomes, enhance healthcare delivery, and enable early intervention in critical situations. In reference [3], a non-invasive test is described for measuring the oxygen saturation level in the blood, providing valuable insights into the effectiveness of oxygen therapy and monitoring respiratory conditions. This test plays a crucial role in evaluating patients' oxygenation status and is widely used in both clinical and home settings. By utilizing a non-invasive sensor, typically a pulse oximeter placed on a patient's fingertip or earlobe, the test measures the percentage of oxygen saturated hemoglobin in the blood. This measurement is essential in assessing oxygen levels and determining the efficacy of oxygen therapy interventions. Healthcare professionals can closely monitor patients' oxygen saturation levels and make informed decisions regarding treatment plans, adjustments in oxygen therapy, or the need for additional interventions. Additionally, the test aids in the early detection of respiratory conditions, allowing for timely interventions and improved patient outcomes.

3 Hardware and Software

3.1 Hardware

In this project we have used myRIO-1900 and ESP8266 device for acquiring the vitals of the patient with the help of medico ECG electrodes and sensor. The myRIO and ESP8266 has the capability to transmit wireless data from the patient to the monitoring system and health supporting staff. The device is optimized to archive accurate measurements at fast sampling rates. Since myRIO has multiple analog input, it is able to acquire varied signals of ECG, EEG, EOG, EMG at a faster rate.

ECG. AD8322 ECG module monitors the electrical activity of patient's heart. This module has three leads which are used to record the ECG of the patient, which are colour coded. The Green lead is used as a reference node yellow and red are used as negative and positive leads respectively to monitor ECG.

EMG. Muscle Bio Amp Candy a candy-size single-channel Electro Myography (EMG) sensor for precise recording of muscle signals with LM324 chip was used to transmits the EMG signals to myRIO in the form of voltage variation from the patients. It is provided with three colour coded lead nodes out of which yellow was use as reference, while red and black were used as positive lead and negative lead respectively.

EOG. BIO Amp Pill with TL084PW chip was used to transmit the EOG signal of the patient to myRIO which is then transmitted to the health monitoring unit or staff. This sensor consists of three colour coded lead wires. Yellow was used as reference while red and black were used as positive and negative node respectively. The reference node is placed behind the left ear while the positive and negative are placed on left temple and right temple respectively.

EEG. BIO Amp Pill with TL084PW chip was used to transmit the EEG signal of the patient to myRIO which is then transmitted to the health monitoring unit or staff. This sensor consists of three colour coded lead wires. Yellow was used as reference while red and black were used as positive and negative node respectively. The reference node is placed behind the left ear while positive and negative are placed on forehead of the patient.

Body Temperature Sensor. For recording temperature of the patient ESP8266 with DS18B20 was used. This sensor is placed on any part of patient's body for recording the body temperature. The ESP8266 supports wireless transfer of temperature data to LabVIEW.

SpO2 and Heartbeat Rate. MAX30102 with ESP8266 is a bio sensor module to record patient's SpO2 (saturation of peripheral blood oxygen) and heartbeat rate (HR) accurately. Left hand middle figure is placed on the sensor for recording patient's SpO2 and HR.

GSM. In this project GSM 900A is connected to ESP8266 which sends alert messages to healthcare professional when there is a sudden change in patient vitals.

Buzzer. The data of the patient is continuously monitored and analysed by LabVIEW if any changed observed an alert is sent to nearby healthcare professional using a Buzzer.

my RIO -1900. In this project NI myRIO 1900 is used as it has 10 analog inputs and 6 analog output and also it can operate autonomously. myRio has 4 different connectors which can be configured to input or output these 4 are:

- In MXP connector A 3 analog input pin are present (AI0AI3) and 2 analog output pin are present (AO0-AO1).
- In MXP connector B 3 analog input pin is present (AI0AI3) and 2 analog output pins are present (AO0-AO1).

- MSP connector C consist of 2 analog input pin (+AI0 AI0, + AI1 -AI1) and 2 output pin those are (+AO0 -AO0, + AO1 -AO1).
- There are two audio jacks which can be configured as input or output.

myRIO also has digital input-output pin on connector A, B and C starting from (DIO1-DIO13) on A and B connector and (DIO0-DIO7) on connector C. myRIO is having UART, accelerometer and input/output audio terminals.NI myRIO is a powerful embedded device that can be used for a wide variety of applications.

ESP 8266. ESP8266 is an affordable Wi-Fi development board which is used as a embedded development board or as a WiFi board for another system or as an autonomous board in IOT. It consists of a 32-bit CPU which is in small size and consumes low power.

3.2 Software

The proposed system uses LabVIEW (2019), a graphical programming language, to transcribe raw data acquired from patients into understandable standard medical parameters. The LabVIEW programming language, also known as G-code, it uses symbols or block instead of standard textual code, which can save time and improve readability of the data. The raw data from the sensors is sent to the LabVIEW G code, where it is processed and transcribed into standard medical parameters.

4 Block Diagram

Figure 1 explains the block diagram of the proposed system. This system comprises of following main components: the sensors, NI myRIO, ESP8266, GSM and the LabVIEW 2019 software. The sensors are placed on the patient at appropriate position and myRIO,

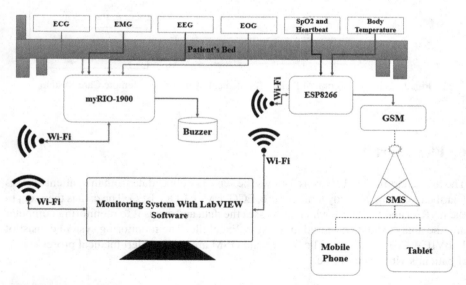

Fig. 1. Block Diagram Consisting Patient Biomedical Monitoring - Remote Care System

ESP8266 and GSM are also placed under the patient's bed. Buzzer is placed outside the patient's room so that it will not disturb the patient.

5 Circuit Diagram

In this circuit diagram the output of ECG, EMG, EOG, EEG sensors are connected to myRIO as analog input. Body Temperature, SpO2 and Heartbeat rate sensor output are connected as input to ESP8266. GSM is connected to ESP8266 as a output device. Buzzer is connected as output device to myRIO. Shown in Fig. 2.

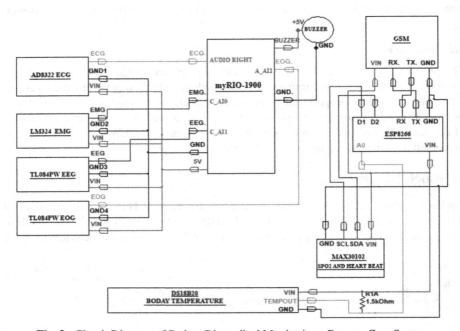

Fig. 2. Circuit Diagram of Patient Biomedical Monitoring - Remote Care System

6 Flowchart

The flow chart in Fig. 3 starts as follows the sensors collect data from the patient such as heartbeat rate, SpO2, body temperature, ECG, EMG, EOG, EEG. The data is then sent to the myRIO and ESP8266 which processes the data and sends it to monitoring computer and the analysed data is stored in a text or Excel file. The monitoring system consist of LabVIEW 2019 software. The buzzer and GSM are used to alert medical professionals if patient's vitals deteriorate.

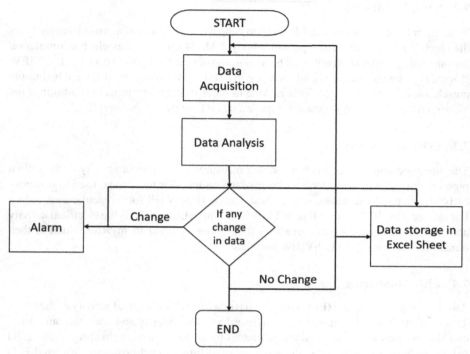

Fig. 3. Flow Chart consisting Patient Biomedical Monitoring - Remote Care System

7 Results

This Patient Biomedical Monitoring - Remote Care System project has the potential to improve patient's status of health and speedy recovery by providing more efficient and effective healthcare. It remotely monitors vital signs of patient. The system utilizes a remote monitoring device to collect data such as the patient's electrocardiogram (ECG), electroencephalogram (EEG), electromyogram (EMG), electrooculogram (EOG), body temperature, and Oxygen saturation (SPO2). These sensors are interfaced with the myRIO-1900, which sends the data to monitoring computer. The LabVIEW software in the monitoring computer will process and analyse the data and stores it as a text or Excel file, if there is any disturbance in the vitals of the patient a SMS will be sent to the health care official and relatives specific to that patient. The device allows early detection of any potential health issues and provide prompt health care needed by the patient.

7.1 ECG Monitoring

To analyse the ECG, we have used AD8322 module which takes the electrical impulse from the patient's body by three lead wires that are reference, positive and negative they are converted into a continues analog values which form the input values to myRIO. myRIO will then transmit the data to LabVIEW software 2019 version.

7.2 EMG Monitoring

To monitor EMG of the patient, Muscle Bio Amp with LM324 chip was used that analyses the electric stimulation of the patient's body. LM324 converts this electric simulation into an analog signal as an output. The analog signals are then sent to myRIO. LabVIEW graphically represents the signals sent by myRIO. EMG is recorded for analyzing the muscle movement of patient. This is done in order to know the muscle condition if the bio signals are low then necessary steps are taken care by the doctors.

7.3 EOG Monitoring

The electrooculogram (EOG) is a test that measures the electrical activity of the retinal pigment epithelium, the outermost layer of the retina. This test can be used to diagnose early-stage macular diseases, such as best disease (Best vitelliform macular dystrophy). For this we used BIO Amp Pill with TL084PW chip which analysis the electrical activity and produce a analog output which is then given as input to myRIO which is then transmitted wirelessly to LabVIEW software.

7.4 EEG Monitoring

An electroencephalogram (EEG) is a test that quantifies the electrical activity of the brain. This data can be used to diagnose brain disorders, like epilepsy and brain tumours. EEG can also be used to assess the damage caused by a head injury. In this project the EEG values are near to accuracy as the we have used limited lead positions. We used BIO Amp Pill with TL084PW which analysis the electrical activity of the brain and produces analog output which is then given as input to myRIO that are trans-mitted wirelessly to LabVIEW software.

7.5 Body Temperature Monitoring

To monitor the body temperature of the patient we have used ESP8266 with DS18B20. DS18B20 sensor sends voltage values to ESP8266 and ESP8266 will calculate the temperature value that is transmitted to LabVIEW using TCP protocol.

7.6 SpO2 and Heartbeat Rate Monitoring

SpO2 and heartbeat rate is monitored using MAX30102 with ESP8266. MAX30102 sensor is used here as it has the capability to calculate both SpO2 and heartbeat rate values at same time. The MAX30102 sensor consist of two LEDs, first emits red light at 660 nm while the second emits infrared light at 880 nm. The sensor transmits both lights that penetrate the skin and measures the reflection with a photo detector which is known as photoplethysmography. This sensor working can be interpreted for both heart rate measurement and blood oxygen saturation measurement.

Heart Rate Measurement. The amount of reflected light changes as the blood flow through the veins of the finger. This creates an oscillating waveform, which can be used to measure the heartbeat rate.

Blood Oxygen Saturation Measurement. The oxygen present in hemoglobin has a specific characteristic, which can absorb infrared light. So that de-oxygenated blood absorbs more red light whereas the oxygenated blood absorbs more infrared light. By measuring the ratio of reflected red and infrared light, the sensor can estimate the blood oxygen level. This method of measuring blood oxygen level is called pulse oximetry. It is a non-invasive and relatively accurate way to measure blood oxygen levels (Figs. 4 and 5).

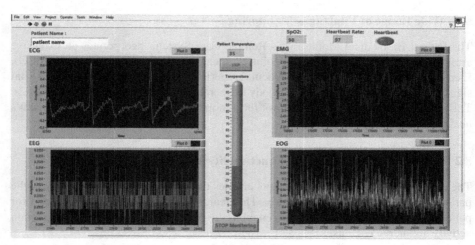

Fig. 4. Patient Biomedical Monitoring - Remote Care System LabVIEW graphical representation of sensors output

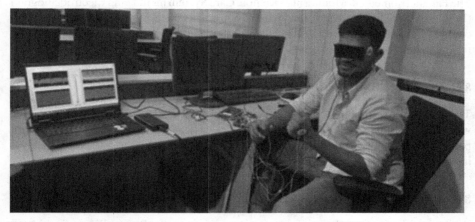

Fig. 5. Patient Biomedical Monitoring - Remote Care System LabVIEW hardware connection with patient

8 Discussion

The Patient Biomedical Monitoring - Remote Care System that we designed will continuously monitor the stable and optimum conditions of blood pressure, SPO2, heartbeat rate, activity of brain, vision, study of nerve functionality, body temperature etc., It is a novel and unique as against the existing patient IoT monitoring system because of its multifaceted functionalities.

8.1 Wide Range of Vital Signs Monitoring

The Patient Biomedical Monitoring - Remote Care System incorporates an extensive array of vital signs monitoring capabilities, including Electroencephalogram (EEG), Electromyogram (EMG), and Electrooculogram (EOG) in addition to the standard vital signs like Electrocardiogram (ECG), body temperature, and Oxygen saturation (SPO2). This comprehensive monitoring allows for a more holistic assessment of the patient's health and neurological status.

8.2 Activity of Brain and Nerve Functionality Monitoring

The inclusion of EEG and EMG sensors enables the system to study the activity of the patient's brain and nerve functionality. This feature can be particularly beneficial for patients with neurological disorders, enabling healthcare professionals to assess brain activity and nerve response.

8.3 Vision Monitoring (EOG)

The Patient Biomedical Monitoring - Remote Care System includes an Electrooculogram (EOG) sensor to track eye movement and vision-related data. Vision monitoring is especially useful for patients with vision-related issues and can provide valuable information for ophthalmological assessments.

8.4 Excel Sheet Database

The system allows for the storage of patient's vital data in the form of Excel sheets, along with their personal details. This feature enables easy accessibility and organization of patient records, making it convenient for healthcare professionals to retrieve and analyse historical data.

8.5 Neuromuscular Activity Control

While the other projects focus on monitoring vital signs, the Patient Biomedical Monitoring - Remote Care System also emphasizes advanced control of neuromuscular activity. This capability can be crucial for neuroprosthetics, rehabilitation, and managing neuromuscular disorders.

8.6 Holistic Patient Monitoring

By combining a wide range of vital signs monitoring with neurological activity assessment, vision monitoring, and neuromuscular control, the Patient Biomedical Monitoring - Remote Care System offers a more comprehensive and holistic approach to patient monitoring and care.

8.7 Potential for Efficient Healthcare

The system's diverse monitoring capabilities and real-time data analysis facilitate early detection of health issues and prompt medical care, improving patient outcomes and overall healthcare efficiency. Overall, the Patient Biomedical Monitoring - Remote Care System stands out for its comprehensive and advanced features, making it a valuable tool for healthcare professionals in diagnosing, monitoring, and managing various health conditions.

9 Deliverable

The Patient Biomedical Monitoring - Remote Care System that is designed will continuously monitor the stable and optimum conditions of blood pressure, SPO2, heart rate, activity of brain, vision, study of nerve functionality, body temperature etc., This Patient Biomedical Monitoring - Remote Care System that we designed is equipped with EOG, ECG, EEG, EMG and body temperature monitoring systems. This equipment is also facilitated with storage of patient's vital data in the form of excel sheets and is available to health care professionals when needed. It allows early detection of any potential health issues and provide prompt health care needed by the patient. Additionally, the system can also give emergency alert to healthcare professionals for sudden anomalies in the patient's vital signs, allowing for swift medical attention.

10 Conclusions

Patient Biomedical Monitoring - Remote Care System provides a better healthcare monitoring of patient's vital data remotely and also alarm the health care officials if there is a sudden change in vitals. myRIO hardware modules receives the data from sensors which is then analyzed by LabVIEW software to obtain graphs of ECG, EEG, EMG, EOG. The processed data is stored in Excels sheet with patients details for future reference.

References

1. Lopes, C.V., Hayes, H., Cullinan, M., Filipe, M.V.: Nanostructured (Ti, Cu) N dry electrodes for advanced control of the neuromuscular activity. IEEE Sens. J. 1 (2023). https://doi.org/10.1109/JSEN.2022.3232264
2. Khan, M.M., Alanazi, T.M., Albraikan, A.A., Almalki, F.A.: IoT-based health monitoring system development and analysis. In: IEEE Volume 2022, Article ID 9639195 (2022), 11 p. (2022)

3. Jegan, R., Nimi, W.S., Jino Ramson, S.R.: Sensors based biomedical frame- work to monitor patient's vital parameters. In: 2020,5th International Conference on Devices Circuits and Systems (ICDCS). IEEE. (2020)
4. Misra, S., Pal, S., Pathak, N., Deb, P.K., Mukherjee, A., Roy, A.: i-AVR: IoT-based ambulatory vitals monitoring and recommender system. IEEE Internet Things J. **10**(12), 1031810325 (2023). https://doi.org/10.1109/JIOT.2023.3238116
5. Hartalkar, A., Kulkarni, V., Nadar, A., Johnraj, J., Kulkarni, R.D.: Design and development of real time patient health monitoring system using Internet of Things. In: 2020 IEEE 1st International Conference for Convergence in Engineering (ICCE), pp. 300–305. Kolkata, India (2020). https://doi.org/10.1109/ICCE50343.2020.9290726
6. Saha, H.N., Paul, D., Chaudhury, Haldar, S., Mukherjee, R.: Internet of Thing based healthcare monitoring system. In: Proceedings of IEEE Annual Information Technology, Electronics and Mobile Communication Conference (IEMCON), pp. 531–535 (2017)
7. Atzori, L., Iera, A., Morabito, G.: The Internet of Things: a survey. Comput. Netw. **54**(15), 2787–2805 (2010)
8. Rathore, D.K., Upmanyu, A., Lulla, D.: Wireless patient health monitoring system. In: Proceedings of IEEE International Conference on Signal Processing and Communication (ICSPC 2013), pp. 415–418 (2014)
9. Gupta, S., Kashaudhan, S., Pandey, D.C.: IOT based Patient Health Monitoring System. Int. Research Jour. of Engg. and Tech. (IRJET) **4**(3), 2316–2319 (2017). (N. Sulaiman, B. S. Ying, M. Mustafa and M. S. Jadin," Offline LabVIEW based EEG)
10. Signals Analysis for Human Stress Monitoring. 2018 9th IEEE Control and System Graduate Research Colloquium (ICSGRC), pp. 126–131. Shah Alam, Malaysia (2018). https://doi.org/10.1109/ICSGRC.2018.8657606

An Energy-Efficient Clustering Approach for Two-Layer IoT Architecture

Annu Malik[✉] and Rashmi Kushwah

Department of Computer Science & Engineering and Information Technology, Jaypee Institute of Information Technology, Noida 201304, U.P., India
annu.knit@gmail.com

Abstract. The Internet of Things (IoT) connects every electronic device to the internet. IoT provides many uses in the daily lives of ordinary individuals, organizations, and society worldwide. It is a rigorous process to communicate with IoT devices to introduce numerous applications to the real-world environment. In this paper, we propose a two-layer IoT architecture paradigm, which comprises an IoT layer and a sensing layer. In applications that require real-time information, sensing devices are crucial parts. To implement IoT-driven applications, a combination of these layers is essential. Proper interaction and utilization of devices and the information received by them across both layers are vital to the successful implementation of any IoT application. In the proposed scheme, the devices are connected in order to facilitate the development of a cluster of sensors at different layers. In addition, clustering facilitates and improves the total endurance of the network. The proposed clustering method is evaluated using the Cooja simulator and compared with the LEACH scheme. The results show that the proposed technique performs better than the LEACH scheme in terms of lifetime of the network, total energy consumption, network coverage rate, and throughput.

Keywords: IoT · Delay · Cluster · Routing · Sensor

1 Introduction

IoT stands for the Internet of Things, a global network of interconnected devices that is used to analyses information with semantics, communication, processing, context, mental processes, and cooperation. The rapid growth of IoT networks has led to advancements in technology as well as applications across a wide range of industries, including automating homes, commercial safety, law enforcement surveillance, and several others. Small micro electric and mechanical devices, or sensor nodes, are used in IoT networking, which connect smart devices and sensor networks worldwide, using a restricted power supply. These developments have increased demand for among machines connections and data and information accessibility at all instances and areas. We present a clustering algorithm for cooperative processing in IoT networks, which enables significant cooperation of large numbers of devices with seamless and reliable communication. IP-enabled devices and underlying devices must be classified and there should be the lowest possible level of communication delay necessary for obtaining data for clustering in the IoT to be effective. The two-layer architectural system is comprised of a top layer of IP-enabled IoT devices and another layer of simple sensor devices as shown in Fig. 1.

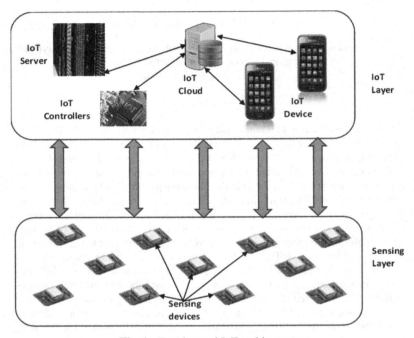

Fig. 1. Two-layered IoT architecture

Clustering methods are utilized to group the devices. Each group in clustering designates a single node as the cluster head, and the cluster head facilitates communication between all of the nodes in the overall network. The technique which we have proposed is novel as it can group devices that are heterogeneous and operate in a layered manner in a dynamic environment of the IoT.

The remaining part of this paper is organized as follows. Section 2 outlines the related work. The design objective and system model are presented in Sect. 3. In Sect. 4, the proposed work is explained. Section 5 displays the implementation and results. Finally, Sect. 6 presents the conclusion.

2 Related Work

In [1], author provided two clustering approaches-heuristic and graph-based clus-tering, respectively, that are best suited for the IoT structure in order to meet all of the demands for minimum transmission and optimum cooperation across the sensors and IoT devices. This strategy promotes a worldwide network setting everywhere. Using the remaining power and the neighbours measure, two meth-ods for clustering have been presented.

According to the requirements of an IoT application, grouping can be done in both top-to-bottom and bottom-to-top di-rections using heuristic and graph-based methodologies. In [2], to maximise the number of groups and cluster heads, three factors are taken into account: The count of neighbouring nodes, the remaining power and range of the node from the primary station. The lifespan attained by the Distributed Hierarchical Agglom-erative Clustering (DHAC) protocol is contrasted with the network lifespan. The DHAC protocol produces substantial energy reductions and extends the lifespan of networks. In [3], by applying the Intelligent Routing Process (IRP), a non-cluster member sends data to the next hop node by balancing the energy needs to send messages to its neighbour against the energy left over from its neighbour. If a group member is chosen as the next hop along the communication route, it denies IRP and sends the messages to the group head, the group head then sends these meassages to the destination. Clusters are not the only tool used for navigation. The Cluster Aided Multipath Routing (CAMP) method produced good results in terms of coverage rate, power usage, and network longevity. In [4], the author suggested a novel clustered routing technique, which focuses on anticipated energy usage. This technique is an economical clustering algorithm which utilises a number of variables. In [5], author presented an adaptive hybrid clustering method that takes into account practical aspects like accessibility, interaction, and transmission in order to facilitate collective computation in In-ternet of Things networks. In this technique every node learns about the CH and builds a cluster by figuring out the highest possible neighbour number with remaining energy. In [6], author presented a method for collecting with mini-mal delay and energy usage by using a matrix filling concept named Delay and Energy-Efficient Data Collection (DEEDC) scheme. This approach ensures that data is possible to gathered in a network with random information which pre-vents time and energy from being wasted by not allocating slots to every cluster and from being acquired in duplicate. The earlier data leads to the proposal of a mixed slot sequencing technique to create an organising scheme that is collision-free, energy and delay-efficient. In [7], author presented a new energy-efficient and data authenticity variant of the Low Energy Adaptive Clustering Hierarchy (LEACH) based routing method using watermarking. There are three stages of routing procedure. We handle energy usage via the cluster generation step in the initial stage. We use the fabrication and placement of watermarks to pro-tect transmission of data in the second stage. Using watermark processing, we verify and identify false information in the third stage of authenticity checking. In [8], author presented Low Energy Adaptive Clustering Hierarchy (LEACH) technique, based on a round cluster head (CH) identification method. There are two primary stages in the scheme, initial set up and stable state. The initial setup and stable state stages of the system incur charges for every round. This dynamic clustering technique distributes energy loads evenly among all nodes within a cluster through self-organization. In [9], author explained the strat-egy for processing data for immersive applications by aggregating the processing capacity of several IoT devices. The framework enables the constant formation of device alliances that pool their unused assets to deliver a service at a pre-determined level of accuracy. In [10], author presented a multi-hop clustering approach to minimise the total amount of necessary web links for IoT systems. The main purpose is to map a group of IoT nodes into the selected group of organisers in order to minimise the entire distance between the nodes and their respective organisers.

3 Design Objective and System Model

The main objective of proposed clustering approach is to minimise communiation delay and overhead by permitting only those nodes that are required to participate in transmission. Along with reduced communication delay for the IoT, it also promises minimal power consumption, maximum network durabil-ity, and a fully adaptable system. A network of N nodes and one sink node are deployed randomly in the IoT environment. A hybrid clustering model is used in which a number of clusters of unequal size and radius are formed. Proactive routing is used for communication in which each node has all the information about its neighbours at any time. Based on the distance of node from sink node, the node can communicate with the sink node in two ways as shown in Fig. 2. First, node can communicate directly with sink node if it is under the trans-mission range of sink. Second, if the node is not in the transmission range of sink, then it must be member of any cluster. All the nodes which can commu-nicate directly to the sink and within defined transmission range called Direct Set (DS) node. A cluster will be formed with a group of nodes called Cluster Members (CM) and a Cluster Head (CH), who are in the transmission range of cluster head. The cluster head is elected on the basis of some parameters that is, Node Degree (ND), Remaining Energy (RE) of node, and Distance from Sink (Dists). The node degree defines the count of nodes or neighbours connected to that node, residual energy is the remaining energy or life of that node, and the distance of node from sink. The node degree and remaining energy or life of elected cluster head should be highest among the members of cluster. Whereas, the distance from sink should be smaller. The load on sensor nodes is decreased because the sink itself handles crucial operations including CH selection. Some assumptions are used, which are given below:

- One base station or sink node is placed in the network which is mobile node.
- N Sensor nodes are deployed randomly and they are fixed.
- Every sensor node uses same energy in transmitting and receiving the packets.
- Each node has its unique ID, connected neighbours count or node degree and a battery which is not replaceable.
- All nodes have a transmission range based on the coverage area and distance from receivers.
- The data received from the neighbour nodes can be aggregated at each node.
- Transmission channel is reliable

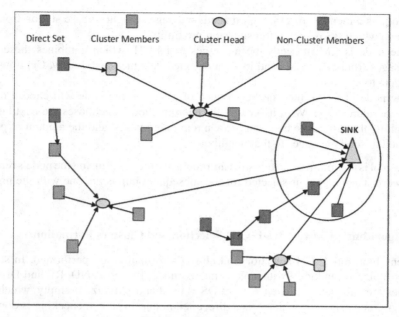

Fig. 2. Routing process

4 Proposed Work: Clustering and Routing Procedures

The proposed scheme uses proactive routing that lowers memory consumption by having every network node only save information about its neighbours at any particular point in time. There are solely three variables in each node's neighbour table (NB-Table):

1. Node Degree (ND)
2. Remaining Energy (RE)
3. Distance from Sink (Dists)

The entire communication area is divided into n zones of the equal area. The CH volume is therefore consistent throughout the area because each zone receives a minimum of one CH. Furthermore, nodes in proposed scheme have no obligation to associate with any CH, in contrast to conventional clustering methods where nodes are obligated to hook onto a single CH. Peering will simply take place if the sensor node and CH both reside within transmission proximity; otherwise, the sensor node will do straight routing. During the clustering, every node updates its NB-Table and shares information with its neighbours. The nodes that make up the DS are those for which base station is a direct neighbour. They exclusively send data to base station directly. After that, base station essentially creates equal-sized zones within the communication area and assigns a CH to each zone. The cluster is made up of all the nodes in a zone that are in transmission proximity of the CH. During routing the communication area is divided into n imaginary zones of similar size, with each node having a distinct zone of origin. Every node should have its NB-Table constructed prior to the start of routing operations. Also, every node is also aware of its status as an NCM, CM, or part of the DS.

1. If Node is a member of DS, then it sends message straight to base station from the node to which BS is within transmission proximity.
2. If the node is CM, it sends the messages to its CH, which combines these with messages from other CMs and its own detected data message. Lastly, CH transmits the data to the sink.
3. If the node is NCM, then in the absence of a cluster, the node will choose multi-hop communication. With its greedy decision method and chooses the next hop by weighing the difference between its own remaining power and the amount of power needed to send messages to that neighbour.

The goal is to minimise the transmission energy while maximising the node's remaining powers. The neighbour selected for the subsequent hop is the one with the highest value.

4.1 Algorithm1: Cluster Head (CH) Selection and Clusters Formation

In given algorithm 1, CH selection and cluster's formation are performed. In step 1, nodes are placed randomly with sink as mobile node. In step 2, ND, RE and Dists for all nodes are calculated. In step 3, a set DS is used and initialize it empty, which will store those node ID's for which S is a direct neighbour. In step 4, checking is done for all nodes u ∈ N, whether S is in transmission range Tx. If yes, then add u in set DS. Repeat steps 4 to 8 for all nodes u. In step 9, check for each node u ∈ N and u / ∈ DS, if yes then store that node IDs which constitutes a cluster member in set CM. CM is initialized empty in step 10. Once CM found then send random messages to find the ND. In step 12, send (ND,RE,Dists) to all neighbours. In step 13, start CH election process. Select a node as CH from all neighbour v of node u with maximum value (ND,RE,Dist). In step 14, Add all neighbours v of u in CM to form a cluster within the range of CH. Repeat steps 10 to 14 for more clusters formation and CH selections. In step 16, update the Cluster formation and CHs.

Algorithm 1 CH selection and clusters formation
1: Input N Number of nodes
2: Input S a Sink node
3: Position all the nodes randomly with sink as mobile node
4: Calculate ND, RE and Dists $\triangleright$ Node degree, Remaining Energy and Distance from Sink
5: Declare DS = ϕ $\triangleright$ Direct Set (DS) to store those node ID's for which S is a direct neighbor
6: for every node u $\in$ N do
7: if S $\in$ T$_x$ then $\triangleright$ T$_x$ Transmission range
8: Add node u to DS
9: end if
10: for each node (u $\in$ N)&(u does not $\in$ DS) do
11: Declare CM = ϕ $\triangleright$ to store node IDs which constitutes a cluster member
12: Send random messages to find the ND
13: Send (ND,RE,Dists) to all neighbors
14: Elect cluster head CH = max(ND$_v$,RE$_v$,Dist$_v$) $\triangleright$ where v is total number of neighbors of u
15: Add v to set CM $\triangleright$ Form a cluster within the range of CH
16: end for
17: Update the Cluster formation and CHs
18: end for

4.2 Algorithm 2: Routing

Algorithm 2 is used to perform the routing procedure. In step 1, ID of the starting node is stored in variable NextNode$_{ID}$. In step 2, check if NextNode$_{ID}$ is equal to Sink ID S$_{ID}$. If yes, then repeat steps 4 to 14 until condition in step 2 is true. In step 5, If node u $\in$ DS, means S is a direct neighbor of u, then u can send packets directly to S. Update NextNode$_{ID}$ as S$_{ID}$. If step 5 is not true then check if u $\in$ CM of any CH then send a request message to CH with ID of u. In step 13, node u receives the broadcast message from CH with MaxSlot value of CH. In step 14, node u sends a packet to CH within the range of MaxSlot value. If node u is neither a DS nor a CM then node u should go to sleep. In step 16, the corresponding CHs sends packets to S continuously and update NextNode$_{ID}$ as S$_{ID}$.

Algorithm 2 Routing
1: Initialize NextNode$_{ID}$
2: while NextNode$_{ID}$ == S$_{ID}$ do
3: if u ∈ DS then
4: Send packets directly to S
5: Update NextNode$_{ID}$ = S$_{ID}$
6: else
7: if u ∈ CM (CH) then
8: Send request message to CH with u$_{ID}$
9: Receive Broadcast message from CH with Max$_{Slot}$[CH]
10: Send Packet to CH within Max$_{Slot}$[CH]
11: else
12: Go to sleep
13: end if
14: CH sends packets to S
15: Update NextNode$_{ID}$= S$_{ID}$
16: end if
17: end while

5 Implementation and Results

To simulate and assess the proposed algorithm, the Contiki-based Cooja simu-lator is used. The Cooja Simulator is run on the Contiki OS 3.0. With a 2.5 GHz processor, Contiki 3.0 requires 2 GB of memory. The simulator provides a choice to select the proper type of mote on the simulator. Using the RPL protocol, the proposed strategy is carried out into practise. Every single IoT device is Tmote sky. The MAC layer protocol for the Internet of Things is IEEE 802.15.4. We have several devices, one remote server, and one gateway deployed. Initially, the dispersed positioning of sensors, gateways, and remote servers builds the network architecture. The analytical indicators such as lifetime of the network, total energy consumption, network coverage rate and throughput are used to measure the performance of our experiments. We have compared the proposed scheme with LEACH [8] scheme.

5.1 Lifetime of the Network

Lifetime of the network can be defined as the time interval that passes between the first node to start sending data and the first node to die.

Here, a network is established by distributing 150 nodes at random around in a 200 m × 200 m area. The clarity and resilience of proposed scheme is demonstrated in Fig. 3. The number of alive nodes with respect to network area shows the liveliness of devices in the corresponding network. As shown in Fig. 3, the lifetime of network for proposed scheme is better than the LEACH [8] technique.

5.2 Total Energy Consumption

Total energy consumption can be defined as the total amount of energy used by all nodes.

$$E_{total} = E_{CH} + E_{CM} + E_{NCM} + E_{DS} + E_{SINK} \qquad (1)$$

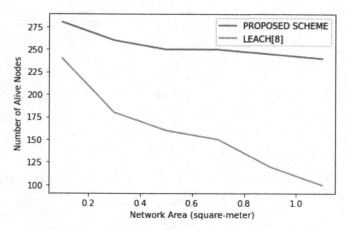

Fig. 3. Number of nodes alive as a function of network area

Here E_{CH}, E_{CM}, E_{NCM}, E_{DS}, and E_{SINK} are the energy consumed by cluster heads, cluster members, non-cluster members, direct sets and sink respectively. E_{total} is the total energy consumed by the system. The comprehensive energy usage of the proposed scheme is shown in Fig. 4. Figure 4 represent the results of proposed scheme and LEACH [8] approach for total energy consumption with respect to the number of rounds. The graph shows that, in comparison to LEACH [8] method, proposed method has a desired energy expenditure curve. Total energy consumption can be defined as the total amount of energy used by all nodes.

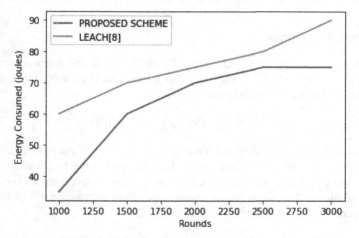

Fig. 4. Total energy consumption of network

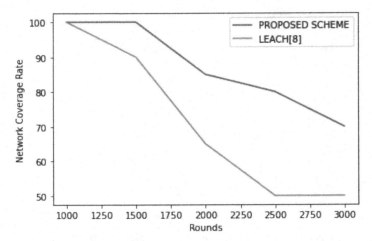

Fig. 5. Network Coverage Rate

5.3 Network Coverage Rate

Network coverage rate is the ratio of the coverage region of the nodes that are ac-tive in that session to the coverage region of all nodes that are active. Compared to its other methods, proposed method retains a higher network coverage rate as an outcome metric. As shown in Fig. 5, compared to LEACH [8] scheme the proposed scheme preserves a higher network coverage rate as an outcome metric. Based on Fig. 5, it is possible to deduced that the coverage rate of proposed scheme remains 100% even after 1500 rounds, while LEACH [8] algorithm have a coverage rate that varies between 90 to 95%.

5.4 Network Throughput

The quantity of data that the system may efficiently obtain and process in a specific period of time is known as its throughput. The throughput is measured in (bits/sec). It can be calculated using given equation:

$$\text{Throughput} = (M_{rcvd} \times M_{length})/\text{Time} \qquad (2)$$

In this case, M_{length} indicates the bit size of each message, Time indicates the period of time required to transmit the data, and M_{rcvd} represents the total number of messages collected effectively. The suggested scheme execution result are shown in Fig. 6. According to the results, the proposed scheme has a higher throughput. The increase in throughput are proportional to rises in device count.

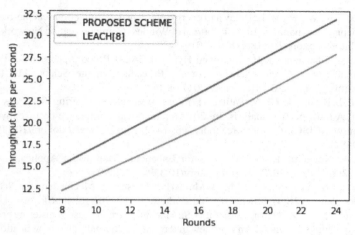

Fig. 6. Network Throughput

6 Conclusion

The biggest challenge in the growth of IoT networks is energy conservation. There are two stages to the algorithm: Clustering and routing. We have devised a highly effective method for sink partitioning the system into multiple areas, along with the clusters. A cluster node sends the data it has sensed to its CH, which then forwards it to the sink. We have developed a straightforward yet sophisticated method that allows non-clustered members to carry out their data using either a straight or clustered plan, based on which method uses least energy. According to simulation results, proposed technique performs better than LEACH scheme in terms of lifetime of the network, total energy consumption, network coverage rate and throughput.

References

1. Kumar, J.S., Zaveri, M.A.: Clustering Approaches for Pragmatic Two-Layer IoT Architecture. Wireless Communications and Mobile Computing **2018**(8739203), 16 (2018). https://doi.org/10.1155/2018/8739203
2. Azad, P., Sharma, V.: Cluster Head Selection in Wireless Sensor Networks under Fuzzy Environment. International Scholarly Research Notices **2019**(909086), 8 (2019). https://doi.org/10.1155/2013/909086
3. Sajwan, M., Gosain, D., Sharma, A.K.: CAMP: cluster aided multi-path rout-ing protocol for wireless sensor networks. Wireless Netw. **25**, 2603–2620 (2019). https://doi.org/10.1007/s11276-018-1689-0
4. Sharad, C., Navpreet, K., Aulakh, I.K.: Evaluation and implementation of clus-ter head selection in WSN using Contiki Cooja simulator. Journal of Statistics and Management Systems **23**(2), 407–418 (2020). https://doi.org/10.1080/09720510.2020.1736324
5. Kumar, J.S., Zaveri, M.A.,: Clustering for collaborative processing in IoT network. In: Proceedings of the Second International Conference on IoT in Urban Space (Urb-IoT '16). Association for Computing Machinery, New York, NY, USA, 95–97,(2016). https://doi.org/10.1145/2962735.2962742

6. Xiang, X., Liu, W., Wang, T.: Delay and energy-efficient data collection scheme-based matrix filling theory for dynamic traffic IoT. Journal of Wireless Communi-cation Network **168**, 1–25 (2019). https://doi.org/10.1186/s13638-019-1490-5
7. Rouissi, N., Gharsellaoui, H.: Improved Hybrid LEACH Based Approach for Pre-serving Secured Integrity in Wireless Sensor Networks. Procedia Computer Science **112**, 1429–1438 (2017). https://doi.org/10.1016/j.procs.2017.08.103
8. Palan, N.G., Barbadekar, B.V., Patil, S.: Low energy adaptive clustering hier-archy (LEACH) protocol: A retrospective analysis, In: 2017 International Conference on Inventive Systems and Control (ICISC), Coimbatore, India, pp. 1–12, (2017). https://doi.org/10.1109/ICISC.2017.8068715
9. Coelho, J., Nogueira, L.: IoT Clusters for Enhancing Multimedia Applications. Sensors (Basel) **22**(23), 9077 (2022). https://doi.org/10.3390/s22239077
10. Yoonyoung, S., Lee, S., Lee, M.: A Multi-Hop Clustering Mechanism for Scalable IoT Networks. Sensors **18**(4), 961 (2018). https://doi.org/10.3390/s18040961
11. Merah, M., Aliouat, Z., Harbi, Y., Batta, M.S.: Machine learning-based clustering protocols for Internet of Things networks: An overview. International Journal of Communication Systems **36**(10), e5487 (2023)
12. Faizan Ullah, M., Imtiaz, J., Maqbool, K.Q.: An Enhanced Three Layer Hybrid Clustering Mechanism for Energy Efficient Routing in IoT. Sensors **19**(4), 829 (2019). https://doi.org/10.3390/s19040829
13. Shahraki, A., Taherkordi, A., Haugen, Ø., Eliassen, F.: A Survey and Future Directions on Clustering: From WSNs to IoT and Modern Networking Paradigms. IEEE Trans. Netw. Serv. Manage. **18**(2), 2242–2274 (2021). https://doi.org/10.1109/TNSM.2020.3035315
14. Rani, S., Ahmed, S.H., Rastogi, R.: Dynamic clustering approach based on wireless sensor networks genetic algorithm for IoT applications. Wireless Netw. **26**, 2307–2316 (2020). https://doi.org/10.1007/s11276-019-02083-7
15. Essalhi, S.E., Raiss El Fenni, M., Chafnaji, H.: A new clustering-based optimised energy approach for fog-enabled IoT networks. IET Networks **12**(4), 155–166 (2023)
16. Roberts, M.K., Ramasamy, P.: An improved high performance clustering based routing protocol for wireless sensor networks in IoT. Telecommun. Syst. **82**, 45–59 (2023). https://doi.org/10.1007/s11235-022-00968-1

Blockchain Research in Healthcare: A Bibliometric Review

S. Neeraj(✉) and Anupam Bhatia

Department of Computer Science and Applications, Chaudhary Ranbir Singh University, Jind, Haryana, India
`neeraj.sahu1202@gmail.com, anupambhatia@crsu.ac.in`

Abstract. The healthcare sector is among the many industries profoundly altered by blockchain technology. The objective of this bibliometric review is to provide a comprehensive analysis of the current state of research concerning the implementation of blockchain technology in the healthcare sector. By conducting a methodical examination of academic publications, we can discern significant patterns, ongoing research topics, and influential figures in this swiftly developing discipline. Our review, which also includes a decentralized system, data security, interoperability, and patient privacy, incorporates an extensive array of topics. This research paper presents a comprehensive analysis of the present state of blockchain investigation within the healthcare industry. It sheds light on potential challenges and future directions by synthesizing findings from a multitude of scholarly sources. Comprehending the trajectory of Blockchain innovation assumes critical importance as the healthcare sector undergoes ongoing digitization and confronts escalating cybersecurity risks. This review makes a significant scholarly contribution by providing insight-rich information that can be utilized by policymakers, practitioners, and researchers who are interested in optimizing healthcare systems through the implementation of Blockchain Technology.

Keywords: Bibliometric · blockchain · VOSviewer · systematic review · RStudio

1 Introduction

Blockchain became popular due to the Bitcoin whitepaper released in October 2008, as it introduced distributed ledger technology [1]. Blockchain technology has been implemented in a genuinely extraordinary number of sectors, including finance, healthcare, and business. The potential of blockchain technology to safeguard patients' sensitive information within the healthcare system is particularly intriguing [2]. Transactions between entities are conducted without the need for a (trusted) third party, as a result of blockchain technology. It functions decentralized transaction validation through the utilization of validators, who are frequently miners. Determining a common understanding among numerous parties who lack mutual trust constitutes distributed consensus. Within the domain of cryptocurrency, this computational dilemma is intertwined with the double-spending dilemma. The latter concerns the verification of a trusted third party (typically a bank) that maintains a ledger of user balances and transactions, to ascertain whether a particular quantity of a digital coin has been expended without its verification. A blockchain is a cryptographic ledger that links a sequence of blocks that have been time-stamped. Securing and immutably, these slabs are airtight [3, 4]. A systematic evaluation is conducted in this paper to evaluate the present state of blockchain research within the healthcare industry. The primary aim of this study is to discern potential healthcare applications of blockchain technology while emphasizing the obstacles and potential avenues for advancement in this domain. After presenting background information, the paper proceeds to provide an elaborate account of the methodology employed. The findings of the analysis comprise a Bibliometric survey, An analysis of the collected data and its characteristics, as well as the findings of a literature quality assessment. The analysis indicates that there is a growing body of research focused on blockchain technology within the healthcare sector. Data exchange, administration of health records, and access control are its principal applications. The results offer significant perspectives on the prospective applications of blockchain technology within the healthcare industry [5]. The preliminary stages of Bitcoin and their pivotal role in establishing blockchain technology are subjects of great intrigue. It's interesting to note that the first iteration of blockchain technology, often referred to as blockchain 1.0, was defined by the initial deployments of digital currencies like Bitcoin. The release of the white paper and the publicly available implementation code were groundbreaking moments that enabled others to build on and improve this technology [6]. The healthcare industry has recently acknowledged blockchain technology as a versatile technology with extensive potential. The potential for blockchain technology to enhance productivity and deliver cost-effective solutions is widely recognized across industries. It facilitates operational efficiency through decentralized and immutable record-keeping, which is validated by community consensus; among its numerous other advantages is the promotion of operational efficiency. In addition, the maintenance of transparent records can promote information symmetry, whereas decentralized organizations can benefit from an accounting, voting, or shareholders registry system based on blockchain technology [3]. Blockchain is a Distributed Ledger Technology that operates through a distributed transactional database, and it is designed to provide cryptographic security and consensus mechanisms [7]. As a result, transactional information is secured and immutable because of the blockchain. Distinguishability, intangibility, and backtracking are attributes that establish blockchain as a disruptive technology [8, 9]. Blockchain enables decentralized transaction validation by employing validators (typically miners) in place of trusted third parties. This obviates the necessity for the participation of a third party in the transaction procedure. By utilizing distributed consensus, which is the capacity of multiple untrusted parties to reach an accord, this is achieved. This computational challenge relates to the double-spending problem within the domain of cryptocurrencies. It concerns how a specific quantity of digital currency can be validated as not having been previously spent, independent of a third-party authority that traditionally maintains records of all transactions and users' balances, such as a bank [5]. Preliminary research has been published on the utilization of blockchain technology across various domains [10]. Zheng et al. [11–13] highlight the architecture and various mechanisms that define blockchain technology. In distinction from the papers previously mentioned, this study offers a methodical examination and assessment of cutting-edge Blockchain Research within the Healthcare domain. The objective of this paper is not only to highlight the present state of blockchain research in healthcare but also to outline the conceivable applications of blockchain in the healthcare sector. Additionally, it aims to identify challenges and potential pathways for future blockchain research.

In this article, a Bibliometric analysis will be undertaken, with a specific emphasis on literature pertaining to the investigation of Blockchain Technology. Its primary objective is to identify the subject matters that have garnered the most scholarly interest in published works. Additionally, the study seeks to identify potential areas for future research in the field of Blockchain Technology. An important contribution to the field has been rendered by the research. The article begins with an examination of the heterogeneous industrial applications of blockchain technology. The findings provide insight into the subject via a comprehensive bibliometric analysis of the literature. Furthermore, the research monitors the patterns of publication about blockchain technology and emphasizes the subjects that garner the most conversation in academic publications. Furthermore, the article uncovers new research avenues in the domain of Blockchain Technology. The paper concludes with the accomplishment of its intended objectives. By employing bibliometric analysis, this study aims to examine publication trends and efficacy in the context of blockchain technology. By utilizing this approach, articles are categorized according to their source and document type, year of publication, subject matter, and the most active source titles; the interrelationships among them are also unveiled. Additionally, the research paper conducts an analysis of clusters by examining the publications' titles, abstracts, and keywords that are present in them.

This paper delves into the field of Blockchain Healthcare and aims to comprehend its intellectual knowledge structure, development trends, and research through bibliometric tools and techniques. To achieve this objective, we conducted an exhaustive bibliometric analysis of 733 published research documents from 2016 to 2023, selected from the Scopus database. The analysis allowed us to gain insights into the knowledge structure and track the progression of Blockchain Technology Research in the Healthcare domain.

2 Background

The foundational principles of blockchain technology are expounded upon in this segment to aid comprehension of the subsequent content.

A peer-to-peer network exchanges data in a decentralized ledger referred to as a blockchain [14]. It successfully resolved a persistent issue referred to as the double-spend problem and was introduced simultaneously with Bitcoin. The process of valid transactions appending to the blockchain in Bitcoin is carried out by the majority of mining nodes, which collectively reach a consensus. Cryptocurrencies were the initial applications of blockchain technology. Utilizing blockchain technology and creating decentralized applications do not, nevertheless, necessitate the introduction of a cryptocurrency.

2.1 Blockchain—Distributed Ledger Technology

The phrase "Blockchain" denotes an assemblage of time-stamped and cryptographically linked blocks. These slabs are sealed securely and impeccably [9]. The content of the previous block is referenced (i.e., a hash value) in each succeeding block as the blockchain grows and obtains more blocks [14]. The entities that make up peer-to-peer (P2P) networks are referred to as nodes on the blockchain. In order for a node to decrypt the messages and access them, a private key is employed. Conversely, for every node in the network to encrypt the messages it receives, a public key is utilized. In order to guarantee the integrity, irreversibility, and non-repudiability of a blockchain, the public key encryption mechanism is thereby integrated. The only prerequisite for deciphering documents encrypted with the corresponding public key is the corresponding private key. The term "asymmetric cryptography" is used to denote this concept. A cryptographic one-way hash function (e.g., SHA256) generates the hash, which serves to establish a link between each node within the blockchain. In addition, it guarantees the anonymity, immutability, and compactness of the block [15]. A signature is appended to each transaction that is completed by a node prior to its distribution to the network for further verification. The implementation of a private key for digital signature purposes serves to facilitate authentication and uphold the integrity of a given transaction. There are two main justifications for this: firstly, the transaction can solely be signed by a user who possesses a distinct private key; and secondly, decryption, which requires the verification of a digital signature, becomes unfeasible in the event of an error that arises during data transmission. Through the utilization of time-stamped blocks, specific nodes, known as miners in the context of particular consensus mechanisms such as proof-of-work or proof-of-stake, organize and encapsulate network-accepted transactions. The process of selecting miners and determining which data is included in the block is governed by the consensus protocol (a consensus protocol is defined in more detail below). Following this, the blocks are distributed throughout the network, where validation nodes utilize the corresponding hash to verify that the block received contains valid transactions and contains a reference to the block before it in the chain. Upon the fulfilment of both prerequisites, the nodes initiate the process of appending the block to the blockchain. In the absence of any of the prerequisites, the block is discarded [5]. The function of network nodes is now addressed. Peer node is the appropriate nomenclature for a node that initiates connections and communications with other nodes in the blockchain network, given that the blockchain is a P2P network. Moving forward, we shall refer to it as a "node" to facilitate understanding. To interact with the blockchain, a user establishes a connection to the blockchain network via a node. All miners are required to operate a completely functional node, which makes the miners mentioned previously a subset of nodes. Therefore, while every node is a miner, not every miner is a node. This situation is documented in a particular form of public blockchain that utilizes the proof-of-work (PoW) consensus mechanism (further explanation to follow in this section). Alternative blockchain networks that employ different forms of distributed consensus, such as proof-of-stake (PoS), obviate the need for mining [7].

A blockchain node is responsible for the following functions:

- Establishing a connection to the blockchain network;
- Maintaining a current ledger of transactions;

- Monitoring and relaying valid transactions;
- awaiting newly sealed blocks;
- validating recently sealed blocks—confirming transactions;
- generating and transmitting new blocks.

2.2 Categories of Blockchains

In general, the categorization of blockchains is predicated on the nature of the data being managed, its accessibility, and the capabilities of the user to execute actions. These consist of:

- Public permissionless,
- Consortium (public permissioned),
- Private.

The Public Permissionless Blockchain, more commonly known as the Public Blockchain, grants the general public visibility and access to its entirety. Strict portions of the blockchain, however, may be encrypted to safeguard the anonymity of participants [11]. Typically, these blockchains are supported by an economic incentive, as in the case of cryptocurrency networks. Illustrative instances of such blockchains comprise Bitcoin, Ethereum, or Litecoin.

The consortium-type blockchain restricts participation in the distributed consensus process to a predetermined set of nodes. It is applicable to a single industry or multiple industries. A consortium blockchain, which is established within a specific industry (e.g., finance), is made accessible to a restricted public and is partially centralized. Conversely, a consortium comprising entities from various sectors (e.g., financial institutions, insurance companies, governmental organizations) enables public access while maintaining a partially centralized trust [11].

A private blockchain restricts network membership to specific nodes only. Consequently, this network is both distributed and centralized. Permissioned networks are private blockchains; they regulate which nodes are permitted to conduct transactions, implement smart contracts, or mine [11]. They are under the management of a single organization that provides trust. It serves a private function. Blockchain platforms such as Ripple [16] and Hyperledger Fabric [17] exclusively provide support for private blockchain networks.

3 Literature Review and Contextual Framework

Blockchain technology has indeed been perceived as a potential facilitator of unique levels of innovation. As a result, researchers have been intrigued by the impacts, functionality, benefits, and challenges that organisations encounter while implementing this technology. The implementation of blockchain technology has been a topic of interest in the academic community for quite some time now. Adams et al. [18] noted the imperative nature of implementing suitable regulations in tandem with technological advancements. It turns out that use cases, practical demonstrations, standards, and lexical coherence are permanent requirements. Taskinsoy [19] stated that blockchain technology has turned over the business and lifestyle sectors. They cited the historical juncture known as the banking crisis to justify the extensive implementation of the measure. Huckle et al. [20] declare that the design feature of blockchain, the peer-to-peer consensus model, enabled us to proactively avert conflicts, as described. However unresolved research obstacles persist, which must be addressed to fortify the advantages of Blockchain Technology. The authors propose that the implementation of a Blockchain consensus protocol that is secure and tailored to the needs of the Internet of Things could potentially result in benefits for the IoT ecosystem as a whole, thereby promoting sustainability [21].

Table 1. Top 7 Articles Frequently Cited on Blockchain Technology

Focus of study	Indexing	Total citations
"A systematic review of current research on blockchain technology" [7]	PubMed	2647
"Execute an organized review of blockchain-based applications, specifying unresolved matters that warrant further investigation and evaluation of their current state and categorization" [8]	ScienceDirect	2182
"Conduct a methodical examination of the obstacles and prospects that arise from the implementation of blockchain technology within the energy industry" [22]	ScienceDirect	2154
"To guide future research and advancements at the intersection of blockchain technology and the Internet of Things, this analysis will examine the obstacles and prospects associated with their integration"[23]	ScienceDirect	1898
"This study aims to examine the application of blockchain distributed ledger technologies in the biomedical and healthcare sectors. It will identify obstacles and prospects, as well as provide recommendations for future research and implementation" [24]	Web of Science	1192
"To obtain a comprehensive understanding of IoT security and to inform future developments in the field, it is imperative to undertake an extensive survey that encompasses application domain analysis, threat identification, and solution architecture evaluation" [25]	IEEE	1191
"Conduct an exhaustive analysis of the blockchain's implementation in the context of the Internet of Things (IoT), examining its potential solutions, obstacles, and uses, to inform future research and implementation strategies"[26]	IEEE	1057

A series of research inquiries to advance our objectives. These are the questions: In recent years, what has been the allocation scheme for blockchain publications? Quantitatively, which blockchain topics are the most frequently published on? Question 3: Which nations make the most significant international contributions to blockchain research? Q4:

What are the current tendencies in blockchain research? The surge in attention towards blockchain technology can be attributed to the widespread adoption of cryptocurrencies and the digital transformation prompted by the global pandemic. The quantity of citations a publication garners determines its level of influence. In order to comprehend the intellectual dynamics of the field, we therefore analyzed the most influential blockchain research publications. In Table 1, the most frequently cited articles from our collection are listed.

Yli-Huumo et al. [7] focused their research primarily on the Bitcoin system and other blockchain applications, such as smart contracts and licensing, according to the most cited articles. Furthermore, the fundamental principles of blockchain technologies within the energy sector were investigated by Andoni et al. [22]. They examined energy applications and adoption cases, including peer-to-peer energy trading, and deliberated on the advantages and disadvantages of utilizing blockchain technology in the energy sector. Kuo et al. [24] directed their attention towards the benefits that blockchain technology offers in the context of healthcare applications. In contrast to conventional distributed databases, they contrasted the most recent biomedical applications and suggested workarounds for potential limitations that may arise during the implementation of blockchain technologies in biomedical care domains. With precise monitoring capabilities, hospitals could ensure the punctual delivery of medications to patients by implementing blockchain technology [24].

4 Methodology

4.1 Search Criteria

Two components constitute the search criteria for this review. The strings "Blockchain" and "Healthcare" comprise the string C1. Conditional 2(C2) comprises a sequence of healthcare-related keywords, including "Doctor", "Records", and "Healthcare". "C1 AND C2" made up the criterion for Boolean expression searches. "Blockchain" AND "Healthcare" AND "Records" AND "Healthcare" OR "Doctor" are examples of search terms that could be employed.

4.2 Criteria for Inclusion and Exclusion:

The selection process for the studies adhered to rigorous criteria for both inclusion and exclusion. To qualify for inclusion, a study needed to satisfy the following criteria:

(I1) The article must have undergone peer review;

(I2) It must have been written in English;

(I3) It must have pertained to the predetermined search terms; and.

(I4) It must have been an empirical research paper, experience report, proof of concept, visionary article, or workshop paper.

Conversely, the following categories of research were omitted from consideration:

(E1) Research articles that failed to provide a distinct emphasis on blockchain technology;

(E2) Articles that did not specifically address the healthcare domain or health records;

(E3) Articles that failed to satisfy all the requirements for inclusion; and.

(E4) Articles consisting of prefaces, keynotes, viewpoints, editorial comments, tutorials, anecdote papers, or presentation slides only.

4.3 Toolkits for Research

Knowledge mapping is an indispensable instrument within the scientific domain, enabling investigators to discern and visually represent concentrations of analogous concepts. Through this process, one can evaluate the advancements made in a specific domain, collaborative scientific efforts, interconnected fields of study, research hubs, and emerging patterns. Researchers can enhance their comprehension of the collective knowledge domain through the utilization of knowledge mapping, which visually represents the outcomes of their inquiries and analyses. This instrument is particularly advantageous in fields of research that are advancing at a rapid rate, such as Blockchain healthcare research, because it facilitates the documentation of knowledge's development and the examination of the expansion of knowledge generation. As of December 4th, 2023, the Bibliometrix R software suite, which can be obtained from biblioshiny package, offers a variety of instruments designed to facilitate quantitative research. The software, which was constructed using the open-source R programming language, provides access to a strong ecosystem as well as efficient statistical algorithms, superior numerical procedures, and integrated data visualization tools. As a quantitative research instrument that is compatible with all main bibliometric analysis techniques, bibliometrics is highly compatible with scientometrics and bibliometrics[27]. The obtained outcomes have been compressed into a CSV file in order to facilitate a bibliometric analysis utilizing VOSviewer. By utilizing co-citation networks, this application facilitates the generation of country-based maps, detects shared keywords across networks, and produces maps comprising a variety of objects[28]. The VOSviewer software enables the clustering, data mining, and mapping of articles extracted from the database.

5 Descriptive Findings

5.1 Chronology of Research on Blockchain in Healthcare

The number of publications concerning blockchain technology in healthcare has evolved, as depicted in the graph in Fig. 1. In 2016, two articles were published in IEEE Access, which ignited initial interest in this subject. The subsequent surge in interest has led to a cumulative count of 733 publications pertaining to the subject matter of blockchain's prospective applications in the healthcare sector. This suggests that blockchain technology in healthcare is regarded as a subject of considerable importance and merits additional investigation by both academic scholars and practitioners. Notably, most articles on this subject have been published in IEEE Access. To summarise, the progression of publications pertaining to blockchain technology in the healthcare sector over time is visually depicted in Fig. 1. The RStudio framework is employed in its development.

As shown in Fig. 2, the number of publications by various authors concerning the application of blockchain technology to healthcare research has increased substantially.

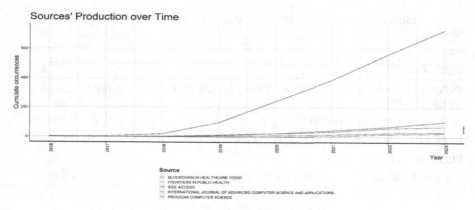

Fig. 1. Sources' Production over Time

The quantity of publications has exhibited a progressive growth parallel to that of the number of authors. By 2023, the number of authors who had published articles had escalated from two in 2016 to thirty. Among these authors, Salah K emerged as the most influential, having authored thirty articles. This exponential expansion is indicative of the increasing scholarly attention towards the application of blockchain technology in healthcare research.

Utilizing the RStudio toolset, Fig. 2 is illustrated.

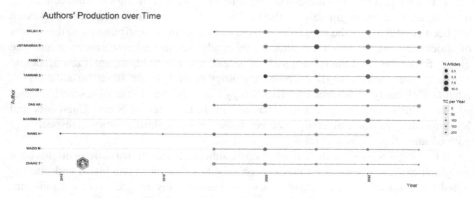

Fig. 2. Author's Production Over Time

5.2 The Key Journals Detail

This research investigates the top six academic journals that, from 2017 to 2023, have published a minimum of 23 articles about healthcare and Blockchain. The number of articles published by the six esteemed journals is detailed in Table 2. According to the study, approximately 82% of the chosen articles are published in these six venues. The

Table 2. Top 6 journals published Articles on Blockchain and Healthcare

Sources	Articles
"IEEE ACCESS"	733
"INTERNATIONAL JOURNAL OF ADVANCED COMPUTER SCIENCE AND APPLICATIONS"	104
"PROCEDIA COMPUTER SCIENCE"	71
"BLOCKCHAIN IN HEALTHCARE TODAY"	34
"FRONTIERS IN PUBLIC HEALTH"	26
"JOURNAL OF MEDICAL SYSTEMS"	23

International Journal of Advanced Computer Science and Applications (IJACS) and Procedia Computer Science are the subsequent most prominent journals in terms of article count, with 733 each. With 104 and 71 articles, respectively, IEEE Access is the foremost journal in its field. 34, 26, and 23 articles, respectively, have been published in Blockchain in Healthcare Today, Frontiers in Public Health, and Journal of Medical Systems. Academic journals specializing in computer science and medicine predominate in the research domain.

5.3 The Distribution of Blockchain Healthcare Research by Region

In recent years, interest has increased regarding the potential for Blockchain technology to bring about a paradigm shift in the healthcare industry. An abundance of research has been conducted on the subject as a consequence of the investigation into the effects of Blockchain technology on healthcare by academicians and researchers around the globe. To comprehend the current state of research in this field, we analyzed academic journal articles concerning Blockchain technology and healthcare. Based on our analysis, Khalifa University of Science and Technology has published the most research articles on this subject, amounting to a remarkable 125. In the interim, Nirma University and Kyungpook National University secured the second and third positions, respectively, with 65 and 48 published articles.

It is critical to specify that Blockchain healthcare research institutions do not have geographical restrictions. This observation signifies that there is a broad and international focus on this subject matter, which is a noteworthy progression as it indicates that the revolutionary capabilities of Blockchain technology in the healthcare industry are acknowledged worldwide. In order to ascertain the significance of this study, we employed the Biblioshiny software to examine the citation frequency of the articles. In the Scopus database, the citation frequency of journal articles was represented by the Local Citation Score (LCS), while the citation frequency of a specific journal article was indicated by the Global Citation Score (GCS). The aforementioned scores offer significant insight regarding the influence of the ongoing research in this field. Collaboration is of the utmost importance in scientific inquiry because it facilitates the exchange of resources and the advancement of knowledge. In pursuit of this aim, we conducted an analysis of the degree of cooperation among the establishments enumerated in

Table 3. Academic institutions engaged in Blockchain-based healthcare research exhibit a high degree of cooperation and knowledge exchange, as evidenced by the institutional collaboration network we analyzed.

The present condition of research regarding the application of Blockchain technology in the healthcare sector is illuminated in a comprehensive manner by our analysis. Blockchain is an area of ongoing research with the potential to revolutionize healthcare, as evidenced by the substantial level of collaboration among academic institutions and the worldwide interest in this subject.

Table 3. Top 10 Institution in the Domain of Blockchain in Healthcare

Affiliation	Articles
"KHALIFA UNIVERSITY OF SCIENCE AND TECHNOLOGY"	125
"NIRMA UNIVERSITY"	65
"KYUNGPOOK NATIONAL UNIVERSITY"	48
"KING SAUD UNIVERSITY"	43
"UNIVERSITY OF MINHO"	35
"KHALIFA UNIVERSITY"	28
"COMSATS UNIVERSITY ISLAMABAD"	25
"LOVELY PROFESSIONAL UNIVERSITY"	23
"SCHOOL OF INFORMATION TECHNOLOGY AND ENGINEERING"	23
"LUT UNIVERSITY"	22

5.4 Knowledge Base of Blockchain Healthcare Research

Citing papers can provide insight into the intellectual underpinnings of a specific field of study within the realm of research. We utilized document co-citation analysis to gain a deeper understanding of the knowledge base and its likely future development in the healthcare blockchain literature. By identifying those two papers that are simultaneously cited by a third paper or multiple papers, we were able to construct a co-citation network and knowledge map. The frequency of reference list citations between two documents is positively correlated with the likelihood that they share a common characteristic. Consequently, cited research papers progressively gain recognition within the scientific community and become associated with a particular scientific paradigm. VOSviewer was used to analyze the co-citation within different papers. The citation network for documents on Blockchain and Healthcare is illustrated in Fig. 3, where each cited document is represented by a node, and the co-citation relationship is denoted by the link between two nodes. Stronger relationships are indicated by broader connections.

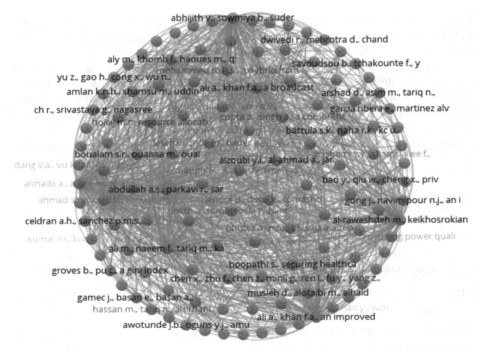

Fig. 3. The Citation Network for Documents About Blockchain and Healthcare

6 Blockchain Healthcare Research Foci

Constant advancements in the sciences generate novel hypotheses, concepts, and methodologies. The vocabulary employed to delineate these emerging phenomena also undergoes a transformation. This is an essential component of the scientific revolution, as it facilitates the systematic investigation of novel domains of inquiry by scientists. Through the analysis of keyword frequency in Blockchain healthcare literature, it is possible to forecast the subjects that will garner the greatest level of interest within a given domain over a specified period of time.

6.1 Keyword Extraction and Frequency Analysis

Keyword analysis is an indispensable tool for discerning significant subjects addressed in the literature and obtaining inroads into burgeoning research domains. We evaluated all selected papers using the VOSviewer Tool for Informatics, extracted the top 10 keywords from 2,689 entries, and prepared them for use in the construction of a keyword co-occurrence network. This network functions as the underpinning for subsequent analysis. In order to maintain precision, we preprocessed the keywords beforehand to eliminate inconsistencies, word compounds, and redundancies. For instance, we excluded "blockchain" and "blockchain technology," as well as "IoT" and "Internet of Things." Fig. 4 depicts the frequency with which the most pertinent words occur.

Blockchain Research in Healthcare: A Bibliometric Review 269

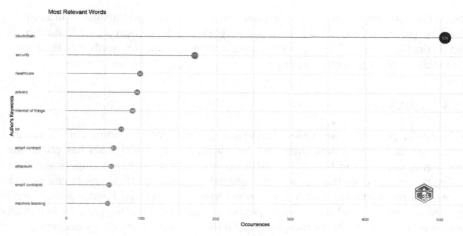

Fig. 4. Most Relevent keywords

6.2 Co-occurrence of Keywords and Research Analysis

As illustrated in Fig. 5, we generated a co-keyword matrix utilizing VOSviewer. The visualization of the keyword co-occurrence network, which is also illustrated in Fig. 5, was derived from the matrix. The nodes on the map symbolize distinct concepts, with their degree of betweenness being denoted by their size. The greater the betweenness of the keyword, the greater the size of the node. Furthermore, in Blockchain healthcare

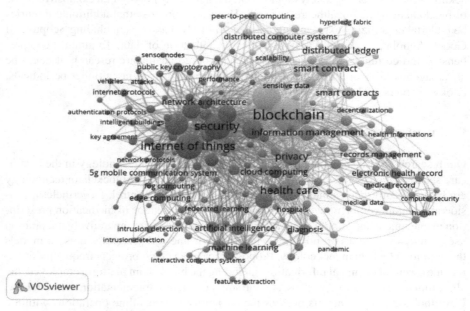

Fig. 5. The visualization of the keyword co-occurrence network

research, the primary emphasis is more likely to be on the keywords positioned in a central position. The interconnections among nodes symbolize the co-occurrence frequency of two distinct keywords. The greater the co-occurrence of the keywords and the closer their relationship, the thicker the edge.

7 Conclusions

A thorough examination was undertaken to ascertain the present condition of blockchain technology implementations within the healthcare sector. Blockchain has numerous potential applications in healthcare, including remote patient monitoring, health data analytics, electronic medical records administration, pharmaceutical supply chains, and biomedical research and education, according to our research. To determine the viability of blockchain technology in healthcare, additional research is required, including an examination of scalability, latency, interoperability, security, and privacy concerns.

8 Limitations

In recent years, there has been a substantial surge in healthcare research about blockchain technology, according to our investigation. Rapidly increasing numbers of academicians are devoting their efforts to this field and robust networks of research collaboration exist between numerous institutions. We propose that to advance this discipline, collaboration among authors from various academic institutions be encouraged. Although keyword analysis can be advantageous for healthcare practitioners and researchers, it is crucial to acknowledge that relying solely on the WOS database may not offer an exhaustive scope of blockchain-related healthcare research. Hence, we propose that additional inquiries be undertaken utilizing extensive and varied scholarly databases, including Scopus and Google Scholar, to compile a more extensive collection of data. To attain a comprehensive and current examination of blockchain-based healthcare research, it could be advantageous to incorporate supplementary materials such as practitioner periodicals, books, chapters, and reports.

9 Future Research Directions

Our research indicates that the implementation of blockchain technology in the healthcare sector is increasing. Currently, the primary applications of blockchain technology are health record management, data exchange, and access control. Nevertheless, considerable untapped potential remains, particularly in the realms of medication prescription management and supply chain administration, which are comparatively uncommon uses. Numerous studies propose novel healthcare frameworks, architectures, or models that employ blockchain technology. Proposals for blockchain projects frequently fail to include essential technical information, including the blockchain platform employed, the algorithm used to reach consensus, and the smart contract implementation methodology. Nevertheless, smart contracts possess the capability to streamline operations within a blockchain platform, thereby expanding their applicability. It may be advantageous to incorporate a prototype implementation or elaborate on implementation details within research proposals. Amidst the extensive implementation of blockchain applications, a multitude of challenges continue to elude resolution. Enhancing the scalability and efficacy of blockchains will bolster their durability. Since the majority of the underlying mechanisms have been widely acknowledged for years, the features they offer are essentially identical when assessed separately. The amalgamation of these attributes, nevertheless, results in numerous industries' intense interest stems from the fact that they are suitable for a multitude of applications. It is anticipated that the domains and sectors in which blockchains find applications will expand beyond the scope of our survey as they mature. Blockchains are not, in fact, a panacea or a viable substitute for databases, contrary to the assertions of numerous individuals. As was previously mentioned, conventional databases are preferable in a variety of circumstances. Furthermore, for each application domain, we have identified the most essential individual characteristics. This improves the process of selecting the appropriate blockchain and the corresponding mechanisms to customize the blockchain according to the specific requirements of the application.

References

1. Nakamoto, S.: Bitcoin: A Peer-to-Peer Electronic Cash System. [Online]. Available: www.bitcoin.org ,last accessed 2023/12/08
2. Abdeen, M.A.R., Ali, T., Khan, Y., Yagoub, M.C.E.: Fusing identity management, HL7 and blockchain into a global healthcare record sharing architecture. Int. J. Adv. Comput. Sci. Appl. **10**(6), 630–636 (2019). https://doi.org/10.14569/ijacsa.2019.0100681
3. Aste, T., Tasca, P., Di Matteo, T.: Blockchain Technologies: The Foreseeable Impact on Society and Industry. Computer **50**(9), 18–28 (2017). https://doi.org/10.1109/MC.2017.3571064
4. Roehrs, A., da Costa, C.A., da Rosa Righi, R.: OmniPHR: A distributed architecture model to integrate personal health records. J Biomed Inform **71**, 70–81 (2017). https://doi.org/10.1016/j.jbi.2017.05.012
5. Hölbl, M., Kompara, M., Kamišalić, A., Zlatolas, L.N.: A systematic review of the use of blockchain in healthcare. Symmetry (Basel) **10**(10), 470 (2018). https://doi.org/10.3390/sym10100470
6. Swan, M.: Blockchain: Blueprint for a new economy. (2015). Last accessed 2023/10/25
7. Yli-Huumo, J., Ko, D., Choi, S., Park, S., Smolander, K.: Where is current research on Blockchain technology? - A systematic review. PLoS One **11**(10), e0163477 (2016). https://doi.org/10.1371/journal.pone.0163477
8. Casino, F., Dasaklis, T.K., Patsakis, C.: A systematic literature review of blockchain-based applications: Current status, classification and open issues. Telematics and Informatics **36**, 55–81 (2019). https://doi.org/10.1016/j.tele.2018.11.006
9. Aste , T., Tasca, P., Di Matteo, T.: Blockchain Technologies: foreseeable impact on industry and society
10. Christidis, K., Devetsikiotis, M.: Blockchains and Smart Contracts for the Internet of Things. IEEE Access, Institute of Electrical and Electronics Engineers Inc **4**, 2292–2303 (2016). https://doi.org/10.1109/ACCESS.2016.2566339
11. Zheng, Z., Xie, S., Dai, H., Chen, X., Wang, H.: An Overview of Blockchain Technology: Architecture, Consensus, and Future Trends, In: Proceedings - 2017 IEEE 6th International Congress on Big Data, BigData Congress 2017, Institute of Electrical and Electronics Engineers Inc., pp. 557–564. (2017). https://doi.org/10.1109/BigDataCongress.2017.85

12. J. H. Park and J. H. Park, "Blockchain security in cloud computing: Use cases, challenges, and solutions," *Symmetry (Basel)*, vol. 9, no. 8, 2017, https://doi.org/10.3390/sym9080164
13. Yin, S., Bao, J., Zhang, Y., Huang, X.: M2M security technology of CPS based on blockchains. Symmetry (Basel) **9**(9), 195 (2017). https://doi.org/10.3390/sym9090193
14. Sleiman, M.D., Lauf, A.P., Yampolskiy, R.: Bitcoin Message: Data Insertion on a Proof-of-Work Cryptocurrency System. In: Proceedings - 2015 International Conference on Cyberworlds, CW 2015, Institute of Electrical and Electronics Engineers Inc., pp. 332–336. (2016). https://doi.org/10.1109/CW.2015.56
15. Dattani, J.. Sheth, H.: Overview of Blockchain Technology. Asian Journal of Convergence in Technology, vol. V Issue I, [Online]. Available: https://dev.to/damcosset/blockchain-what-is-in-a-block-48jo
16. Ripple: Ripple. Ripple—One Frictionless Experience To Send Money Globally. (2018)
17. Androulaki, E., et al.: Hyperledger Fabric: A Distributed Operating System for Permissioned Blockchains. In: Proceedings of the 13th EuroSys Conference, EuroSys 2018, Association for Computing Machinery, Inc, (2018). https://doi.org/10.1145/3190508.3190538
18. Adams, R., Parry, G., Godsiff, P., Ward, P.: The future of money and further applications of the blockchain. Strateg. Chang. **26**(5), 417–422 (2017). https://doi.org/10.1002/jsc.2141
19. Taskinsoy, J.: Blockchain: A Misunderstood Digital Revolution. Things You Need to Know about Blockchain
20. Huckle, S., Bhattacharya, R., White, M., Beloff, N.: Internet of Things, Blockchain and Shared Economy Applications. Procedia Computer Science **98**, 461–466 (2016). https://doi.org/10.1016/j.procs.2016.09.074
21. Makhdoom, I., Abolhasan, M., Ni, W.: Blockchain for IoT: The Challenges and A Way Forward
22. Andoni, M., et al.: Blockchain technology in the energy sector: A systematic review of challenges and opportunities. Renewable and Sustainable Energy Reviews **100**, 143–174 (2019). https://doi.org/10.1016/j.rser.2018.10.014
23. Reyna, A., Martín, C., Chen, J., Soler, E., Díaz, M.: On blockchain and its integration with IoT. Challenges and opportunities. Futur. Gener. Comput. Syst. **88**, 173–190 (2018). https://doi.org/10.1016/j.future.2018.05.046
24. Kuo, T.T., Kim, H.E., Ohno-Machado, L.: Blockchain distributed ledger technologies for biomedical and health care applications. Journal of the American Medical Informatics Association **24**(6), 1211–1220 (2017). https://doi.org/10.1093/jamia/ocx068
25. Hassija, V., Chamola, V., Saxena, V., Jain, D., Goyal, P., Sikdar, B.: A Survey on IoT Security: Application Areas, Security Threats, and Solution Architectures. IEEE Access, Institute of Electrical and Electronics Engineers Inc **7**, 82721–82743 (2019). https://doi.org/10.1109/ACCESS.2019.2924045
26. Fernández-Caramés, T.M., Fraga-Lamas, P.: A Review on the Use of Blockchain for the Internet of Things. IEEE Acces, Institute of Electrical and Electronics Engineers Inc **6**, 32979–33001 (2018). https://doi.org/10.1109/ACCESS.2018.2842685
27. Kemeç, A., Altınay, A.T.: Sustainable Energy Research Trend: A Bibliometric Analysis Using VOSviewer, RStudio Bibliometrix, and CiteSpace Software Tools. Sustainability (Switzerland) **15**(4), 3618 (2023). https://doi.org/10.3390/su15043618
28. Kuzior, A., Sira, M.: A Bibliometric Analysis of Blockchain Technology Research Using VOSviewer. Sustainability (Switzerland) **14**(13), 8206 (2022). https://doi.org/10.3390/su14138206

Enhancing IoT Security Through Hierarchical Message-Passing Graph Neural Networks: A Trust-Driven Strategy for Identifying Malicious Nodes

C. Senthil Kumar[✉] and R. Vijay Anand

School of Computer Science Engineering & Information Systems, Vellore Institute of Technology, Vellore, India
{senthilkumar.c,vijayanand.r}@vit.ac.in

Abstract. Internet of Things (IoT) gadgets connect and simplify life in several ways. The independence of these technologies poses some issues, such as the protection of security and privacy from malicious and compromised nodes inside the network. To overcome this complication, Enhancing IoT Security Through Hierarchical Message-Passing Graph Neural Networks: A Trust-Driven Strategy for Identifying Malicious Nodes (HMPGNN-Trust-IoT) is proposed. These issues are addressed by the proposed HMPGNN-Trust-IoT solution, which develops a mechanism for recognizing malicious and compromised nodes utilizing trust-based deep learning. With the help of the dataset provided, HMPGNN-Trust develops a worldwide model to predict the IoT node's irrational behavior. A global dataset comprising 19 trust parameters on three main categories including knowledge, experience, and reputation are used in the proposed HMPGNN-Trust-IoT technique. The data are fed to pre-processing. In preprocessing, it cleans and normalizes the data in the dataset by utilizing a Nanoplasmonic Ultra Wideband Band Pass Filter (NUWBF). HMPGNN-Trust uses the idea of communities using dedicated servers to divide the dataset into smaller portions for effective training to lessen the computational strain. Finally, it classifies the nodes by utilizing a Hierarchical Message-Passing Graph Neural Network (HMPGNN) into benign and malicious nodes. The proposed HMPGNN-Trust-IoT is implemented using Python. To identify Trust Management in IOT performance metrics like precision, accuracy, and F1-score are considered. Performance of the HMPGNN-Trust-IoT approach attains 24.11%, 28.56% and 22.73% high accuracy, 21.89%, 23.04%, and 9.51% high precision, 25.289%, 15.35%, and 19.91% higher F1-Score compared with existing methods such as the Securing IoT along Deep Federated Learning: A Trust-based Malicious Node Identification method (TM-ANN -IoT), the Trust-driven reinforcement selection technique in federated learning on Internet of Things devices (TM-SDQN-IoT) and the role of machine learning methods in internet of things-based cloud applications (TM-SVM-IoT) models respectively.

Keywords: Internet of Things (IoT) · Nanoplasmonic Ultra Wideband Bandpass Filter (NUWBF) · Hierarchical Message-Passing Graph Neural Network (HMPGNN) · Trust Management · trust dataset · benign and malicious nodes

---

© The Author(s), under exclusive license to Springer Nature Switzerland AG 2025
M. Malhotra (Ed.): IETCIT 2024, CCIS 2125, pp. 273–285, 2025.
https://doi.org/10.1007/978-3-031-80839-5_21

1 Introduction

A new idea called the Internet of Things (IoT) envisions connecting common devices to the Internet. These devices distribute information through the Internet due to their interconnection. This has led to an abundance of new services and applications that are changing the way it works and lives. However, these devices' autonomous connectivity creates dangers to the safety and privacy of produced data. It is significant to find solutions to pressing problems, such as protection, sensitive data protection, and communication integrity among devices to ensure the secure and IoT network's reliable operation. Trust management is crucial for securing IoT networks because it ensures the legitimacy and dependability of the participating devices and information. Trust management allows for a safe channel of communication between devices and shields malicious devices from endangering network security. The main problem of trust management in IoT networks is rogue node identification, which can undermine network security by carrying out malicious actions such as data tampering, eavesdropping, and denial-of-service assaults. Many trust management techniques have been offered as solutions to these research issues including reputation-based systems, cryptography, and machine learning-based methods. Aspects of Blockchain technology and federated learning are frequently present in most of the available techniques for maintaining trust in IoT contexts. Numerous nodes can collaborate to create a machine-learning model through federated learning without disclosing any of their personal information. This is done by employing locally aggregated gradients during model training, which are combined to update a common model. This technique boosts the model's effectiveness and privacy while also ensuring that it is trained on a wider variety of data. Additionally, blockchain technologies provide an untouchable and secure ledger of all network transactions. In an IoT environment, blockchain can be used to record every device interaction, including the execution of smart contracts, and asset transfer.

The main contributions of this research are summarized below.

- In this research, HMPGNN-Trust-IoT is proposed.
- Based on IoT trust management, the suggested technique uses a Hierarchical Message-Passing Graph Neural Network (HMPGNN) to recognize malicious and compromised nodes for detecting malicious and compromised nodes.
- The proposed method utilizes a novel global dataset made up of trust parameters to forecast nodes' aberrant behavior to overcome shortcomings of conventional trust management strategies in the Internet of Things, which is susceptible to attacks like whitewashing and badmouthing.
- The proposed technique also eliminates significant capacities needed for training purposes.

The remaining manuscripts are organized as follows: Section 2 reviews the literature survey, Section 3 describes the proposed methodology, Section 4 proves the results, and Section 5 concludes the manuscript.

2 Literature Survey

Numerous research works were presented in the literature related to deep learning depend Trust Management in IOT, a few recent works are reviewed here,

In 2023, Awan, K.A, et.al, [6] presented Securing the Internet of Things through Deep Federated Learning: A Trust-base Malicious Node Identification Method. A global dataset provided the input. The presented method makes use of a new trust dataset comprised of 19 variables drawn on knowledge, experience, and reputation were three main components. To reduce computational load, Fed Trust uses a community concept with dedicated servers to split datasets into smaller sections for successful training. The presented method was thoroughly compared to alternative methods including accuracy, and precision measures to examine the performance of the Internet of Things network. It provides high accuracy with low precision.

In 2022, Rjoub, G. et al. [7] presented the Trust-driven reinforcement selection technique in federated learning on Internet of Things strategies. The input was given by a global dataset. It introduces DDQN-Trust, a double deep Q learning-based selection technique that considers the energy levels and trust ratings of IoT devices to identify the most effective scheduling alternatives. Finally, it includes solutions for the FedAvg, FedProx, FedShare, and FedSGD federated learning aggregation methods. Reality world dataset tests demonstrate DDQN-Trust solution consistently beats the two primary benchmarks of the DQN and random scheduling methods. It provides low F1-Score and high precision.

In 2022, Mishra, S, et.al [8] presented the role of machine learning methods in IoT basis cloud applications. The input was given by the dataset. To ensure that these connected (smart) devices continuously learn from the available data sets and improve themselves without any manual intervention, enormous amounts of data were evaluated and processed from all of these connected (smart) devices. The foundation for machine learning is created at this point. It provides low accuracy and a high F1 score.

In 2023, Arshad, Q.U.A, et.al, [9] have presented Blockchain-based decentralized trust management in IoT: schemes, necessities, and difficulties. Here, the input was given by the dataset. The presented report provides a comprehensive survey of Blockchain-based decentralized trust management systems in the Internet of Things for analysis and evaluation. The presented paper makes three contributions. First, it presents a thorough comparison of the state-of-the-art BCDTMS designed in several IoT classes, including the Internet of Medical Things, Vehicles, Industrial IoT, and Social IoT. The presented technique conducts a thorough analysis of current BCDTMS in the literature of Blockchain areas and TM to make the study comprehensive. It provides low precision and high accuracy.

In 2023, G., Bentahar, et.al, [10] presented that the Explainable Trust-aware Selection in Autonomous Vehicles utilizing LIME for One-Shot Federated Learning. The dataset provides the input. The selection of dependable autonomous vehicles (AVs) for training was essential for the effectiveness and dependability of federated learning systems. The presented study offers a one-shot federated learning strategy for AV selection that considers the performance of local interpretable model-agnostic explanations technique and the performance of each AV. The Trust-Aware XAI LIME Deep Q-learning-based AV selection technique was created by modifying the original XAI LIME Deep Q-learning-based AV selection technique to incorporate the trust metric. It provides a high F1-Score and low precision.

3 Proposed Methodology

HMPGNN-Trust-IoT is discussed in this section. This section presents a clear description of the research methodology used for trust management in IoT from a Global dataset. The block diagram of HMPGNN-Trust-IoT is represented in Fig. 1. Thus, the detailed description of HMPGNN-Trust-IoT is given below,

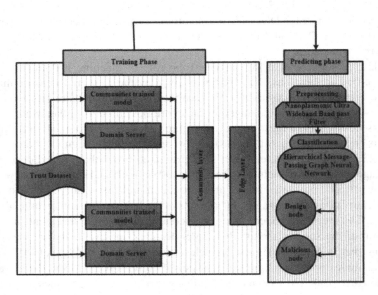

Fig. 1. Block diagram of the proposed HMPGNN-Trust-IoT Trust Management in IoT

3.1 Data Acquisition

Initially, the data were taken from the global Database [13]. The dataset for the suggested approach is made up of various trust component attributes; these elements are chosen to give nodes the intelligence to anticipate the behavior and the patterns of various IoT potential assaults to sustain trustworthiness. There are several observations for each of the 90788 parameter values in the dataset.

3.2 Trust Management Based on HMPGNN-Trust

The suggested Trust solution makes use of the federated learning and community-based parading concept to maintain trustworthiness. Edge nodes with modified federated learning implementation architecture were used to train the model according to the stated approach. The proposed HMPGNN-Trust-IoT approach architecture, dataset characteristics, dataset partitioning, and deep-federated learning training will all be covered in more detail in this section. In this approach, edge nodes are exempted from data exchange while they train their models for federated learning. This choice, unlike some others, ensures that no private information will be shared. Large-scale IoT networks are a good fit for the suggested approach because of the capacity of deep federated learning to share the computational load across edge nodes. When compared to traditional, centralized schooling structures, this is very time and money-efficient. Protection against outside forces: The suggested method is more resistant to common attacks like whitewashing, and badmouthing to innovative trust datasets with criteria. It shows how Trust is achieved in comparison with other solutions when being attacked in various ways. Due to its capacity, this method is applicable in dynamic and heterogeneous Internet of Things networks.

The federated learning idea has changed. The domains and communities are utilized to train edge nodes using tiny portions of the information to improve efficiency and training length. The proposed Trust technique uses HMPGNN-Trust to train the model. The algorithm shows how the suggested mechanism functions in its entirety.

Algorithm: Trust Management module Training Workflow
Input: Global dataset $(Trust_{Data})$
Output: Global Model $(Trust_{Midel})$
Procedure Environment Creation
Create domains: $D_p (p = 1,2,....m)$
Create dedicated Servers: $d_s = d = 1,2,...m$
Domain communities: $C_{pq} (p = 1,2,....m)$
Communities edge nodes: $F_{pq} = f = 1,2,...m$
Procedure Dataset Splitting & Allocation
Split $Trust_{Data}$ into n parts: $T_p = T/m$
$Trust_{Data}$ allocation, $R_p \leftarrow T_p$
Edge node Selection: $F_{pq} : CR_{pq}$, for training: $F_{pq}(R_{pq} > threshold)$
Split T_p into chunks: $T_p = t_{pq}$ where $t_{pq} = T_p / m$
Allocate every dataset chunk: to selected edge nodes, $F_{pq} : F_{pq} \leftarrow t_{pq}$
Procedure Training and Communities Trained Model Formulation
Training of edge nodes: to formulate a local model, $L_{pq} : L_{pq} = Train(F_{pq}, t_{pq})$
Local trained models transmission: to domain server, $R_p : R_p \leftarrow L_{pq}$
Merging of locally trained models: to formulate a domain model, $DL_p : DL_p = Mer(F_{pq})$ Transmission of domain model: to central server, $ER \leftarrow DL_p$
Procedure Global Model Formulation
For each domain: p, receive all the community-trained models, TL_q
Procedure Global Model Transmission
Transmitted global model: to domain servers, $R_p : R_p \leftarrow Trust_{Midel}$
Received global mode: by the central server ER
For every community node: in the community do
Update F_{pq} with $Trust_{Midel} : F_{pq} \leftarrow Trust_{Midel}$
Confirmation from edge nodes: have acquired an updated global model, $Trust_{Midel}$
Exit

3.3 Preprocessing Using Nanoplasmonic Ultra-Wideband Bandpass Filter (NUWBF)

In this step, NUWBF performs the data pre-processing [14] which is utilized for cleaning and normalizing the data in the dataset. Typically, cascaded resonators are used to create filters, which are then synthesized based on predetermined specifications, such as center frequency, bandwidth, and insertion loss for a bandpass filter. It is expressed in Eq. (1),

$$\frac{\vartheta^!}{\vartheta_1^!} = \frac{1}{\vartheta}\left(\frac{\vartheta}{\vartheta_0} - \frac{\vartheta_0}{\vartheta}\right), \tag{1}$$

where and. With the use of the ladder network components b_1, and b_5, , as listed the 5th-order prototype LPF may be created. A capacitor (b_1), an inductor (b_2), a capacitor (b_3), an inductor (b_4), and a capacitor (b_5) make up this circuit which is evaluated in Eq. (2) and (3),

$$b_0 = b_4 = b_6 = 1, \quad b_n = 4 \times \left(\frac{w_{n-1} \times w_n}{u_{n-1} \times b_{n-1}}\right), \tag{2}$$

where, $n = 2, 3, ..., k$

$$w_n = \sin\left(\frac{(2n-1)\gamma}{2k}\right), \quad u_n = \beta^2 + \sin^2\left(\frac{n\gamma}{k}\right) \tag{3}$$

The coupling structure in this instance is a configuration of asymmetrically connected lines. A 3-wire TL structure was taken to describe the coupling gap among these asymmetrically coupled lines. The characteristic impedance (I_{in}) of 3-wire TL linked using the 3-wire TL model provided by SRR using the ring-coupled line shown in Eq. (4)

$$I_{in} = I_0 \frac{I_d + iI_0 \tan(\lambda f)}{I_0 + iI_d \tan(\lambda f)}, \tag{4}$$

where, I_0 represent open stub's load and input impedances. In the next subsections, the phase constant of the open stub is described. A UWB bandpass filter with three and five resonators is created to enhance pass-band performance and denial capabilities. The categorization procedure is then applied to the pre-processed data.

3.4 Classification Using Hierarchical Message-Passing Graph Neural Network (HMPGNN)

In this section, classification using HMPGNN [15] is discussed. The fundamental concept is to create a hierarchical construction that recognizes inventive intra- and inter-level propagation techniques to restructure each node in flat to multiple-level super graphs. Meso and macro-level semantics are included in the learned node representations in the produced hierarchy. After collecting node representations with flat information aggregation, it performs bottom-up propagation utilizing node representations to update node representations in H_2, H_p hierarchy H in Eq. (5)

$$m_{p_a^g}^{(k)} = \frac{1}{|p_a^g|+1}\left(\sum_{p^{g-1}\in p_a^g} H_{p^{g-1}}^{(k)} + H_{p_a^g}^{(k-1)}\right) \quad (5)$$

where denotes super node and signifies node that is a member of signifies some nodes of level $k-1$ that are a part of the super node p_a^g, $|p_a^g|+1$ denotes node representation of st I created by layer 1 in the graph $|p_a^g|$. $m_{p_a^g}^{(k)}$ Denotes representation of st I updated. Internal propagation to transport information throughout each level's graph $(H_1, H_2..., H_p)$, investigates the standard flat GNN encoders and it is expressed in Eq. (6)

$$\begin{aligned}a_s^{(k)} &= AGG^P\left(\left\{\hat{S}_{pq}^g, s_p^{(k)} | p \in M^g(q)\right\}\right),\\ a_r^{(k)} &= AGG^P\left(\left\{\hat{S}_{pq}^g | p \in M^g(q)\right\}\right)s_p^{(k)},\\ h_r^{(k)} &= COM\left(a_s^{(k)}, a_r^{(k)}\right)\end{aligned} \quad (6)$$

where $s_p^{(k)}$ denotes node set next to v at level, $h_r^{(k)}$ signifies aggregated node representation v based on neighborhood data locally, $a_r^{(k)}$ signifies node representation u following bottom-up propagation at p-th layer. To update the original node G representation, it employs node representations in, etc. For other jobs, the value of communications at various levels may vary. Therefore, it utilizes the attention mechanism described to learn of various level contribution weights during top-down integration in an adaptive manner and it is expressed in Eq. (7)

$$m_r^{(k)} = \text{Re } OS\left(M.mean\left\{\kappa_{pq}h_r^{(k)}\right\}\right)\forall p \in E(q) \cup \{q\} \quad (7)$$

where κ_{pq} signifies the set of super nodes at various levels 2, K to which node v belongs $|E(q)| = s-1$, ReLU denotes activation function, $_{pq}$ signifies trainable normalized attention coefficient connects node v to the super node u or itself. The layer $h_r^{(k)}$ has a created node representation called $\kappa_{pq}h_r^{(k)}$.. The last layer's (L) output node representations are produced using Eq. (8) as follows,

$$o_p = \psi\left(M.mean\left\{\kappa_{pq}h_r^{(k)}\right\}\right)\forall p \in E(q) \cup \{q\} \quad (8)$$

where, to transform values into the range [0, 1], ψ is the Euclidean normalization function. The Trust procedure starts with the construction of domains and the distribution of the global dataset to n parts, with every part distributed to the specific server via constructing n domains. A tiny portion of the dataset is allotted to each edge node by the domain server, which has powerful computing capabilities. The provided dataset is then used to start training the local model on the chosen edge nodes. The local model is sent to the appropriate domain server after being created locally. The domain-trained mode, which was created by merging these trained local edge node models on each domain server was then created. The central server receives the domain-trained model for merging, where it is combined with other domain-trained models and shared with the edge node to create a global model that can make predictions. Finally, HMPGNN detects the Trust Management in IOT as benign and malicious nodes.

4 Result with Discussion

The experimental outcomes of the suggested method are discussed in this section. The simulations are performed in Jupyter Notebook utilizing Python language. The suggested method is then simulated in Python using the mentioned performance indicators. The proposed TM-HMPGNN-IoT approach is implemented in Python using a historical dataset. The obtained outcome of the proposed TM-HMPGNN-IoT approach is analyzed with existing systems like TM-ANN-IoT [6], TM-SDQN –IoT [7], and TM-SVM –IoT [8] models respectively.

4.1 Trustworthiness Assessment Metrics

Figure 2 represents the results. The proposed HMPGNN-Trust outperforms other methods TM-ANN-IoT, TM-SDQN -IoT, and TM-SVM -IoT with a 96% higher accuracy, 92% higher Precision, 94% higher F1-Score Trust-based intrusion identification in the Internet of Things context is improved significantly. The outcomes proposed the use of the HMPGNN-Trust model to increase intrusion identification, point to a promising method for safeguarding IoT networks.

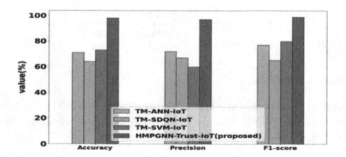

Fig. 2. Performance Analysis of HMPGNN- Trust

4.2 Classification Performance Measures

This is a crucial step for choosing the optimal classifier. Performance measures are assessed to assess performance, including accuracy, precision, and F1-Score. To scale performance metrics, the performance metric is deemed. True Negative, True Positive, False Negative and False Positive values are needed to scale performance metrics.

- True Negative (TN): Count appropriately categorized as a benign node.
- True Positive (TP): Count of appropriately categorized as malicious nodes.
- False Positive (FP): Count of benign nodes categorized as malicious.
- False Negative (FN): Count of malicious nodes categorized as benign.

4.2.1 Accuracy

The proportion of correctly identified data points to all the data points in the dataset is used to measure accuracy and measured by Eq. (9),

$$Accuracy = \frac{TP + TN}{TP + TN + FP + FN} \qquad (9)$$

4.2.2 Precision

The percentage of actual positive cases among all instances that were predicted to be positive and computed by Eq. (10)

$$Precision = \frac{TP}{TP + FP} \qquad (10)$$

4.2.3 F1-Score

The harmonic mean of precision and recall is used to calculate the F1 Score, which is a metric for a model's accuracy and is calculated by Eq. (11)

$$F - Score = 2 * \frac{Precision * recall}{Precision + recall} \qquad (11)$$

4.3 Performance Analysis

Figures 3–5 depicts the simulation results of the proposed HMPGNN-Trust-IoT method. Then, the proposed HMPGNN-Trust-IoT method is likened to existing TM-ANN-IoT, TM-SDQN-IoT, and TM-SVM -IoT methods respectively.

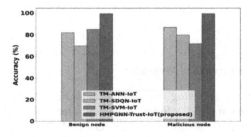

Fig. 3. Performance Analysis of Accuracy

Figure 3 displays the Accuracy analysis. The proposed HMPGNN-Trust-IoT method attains 87%, 78%, and 80% higher accuracy for benign nodes: 31%, 33%, and 26% higher accuracy for malicious nodes estimated to the existing TM-ANN-IoT, TM-SDQN -IoT, and TM-SVM -IoT models.

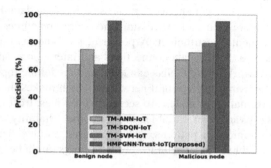

Fig. 4. Precision analysis

Figure 4 shows precision analysis. The proposed HMPGNN-Trust-IoT attains 28%, 30%, and 25% higher precision for benign nodes; 22%, 31%, and 30% higher precision for malicious nodes estimated to the existing TM-ANN-IoT, TM-SDQN -IoT, and TM-SVM -IoT models.

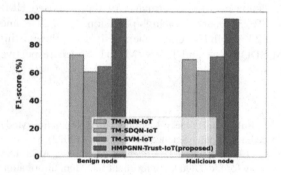

Fig. 5. Performance of F1-Score Analysis

Figure 5 displays the performance of the F1-score analysis. The proposed HMPGNN-Trust-IoT method attains 26%, 22%, and 29% higher F-scores for benign nodes; 23%, 28%, and 26% higher F-scores for malicious nodes estimated to the existing TM-ANN-IoT, TM-SDQN -IoT and TM-SVM -IoT models.

5 Discussion

A novel HMPGNN-Trust-IoT model to identify the benign and malicious nodes in IoT from the Global Database is developed in this paper. IoT devices frequently lack resources, making them susceptible to a variety of security risks. Malicious nodes can attack other devices, interfere with network processes, and steal data. Because IoT environments have a limited amount of processing and energy resources, traditional security methods are frequently insufficient. A powerful machine learning method called HMPGNNs can analyze the intricate connections and interactions among nodes in an Internet of Things network. H-MPGNNs can learn and model the topology and behavior of the network by using graph neural networks, which makes them well-suited for spotting anomalies or malicious nodes. To secure IoT devices, trust is essential. Trust-driven solutions use a variety of variables to assess the reliability of network nodes. Node behavior, communication patterns, device features, and historical data are a few examples of these variables. Nodes with lower trust ratings might be suspected of being malevolent.

6 Conclusion

In this section Enhancing IoT Security Through Hierarchical Message-Passing Graph Neural Networks: A Trust-Driven Strategy for Identifying Malicious Nodes (HMPGNN-Trust-IoT) was successfully implemented for classifying the benign and malicious nodes. The proposed HMPGNN-Trust-IoT approach is implemented in Python utilizing the dataset of Global Database. The performance of the proposed HMPGNN-Trust-IoT approach contains 28%, 31%, and 25% high precision, 27%, 24%, and 22% high accuracy, 21%, 26%, and 27% high F1-score, when analyzed to the existing method such as TM-ANN-IoT, TM-SDQN -IoT and TM-SVM -IoT models respectively.

References

1. Bangui, H., Ge, M., Buhnova, B.: Deep-Learning based Reputation Model for Indirect Trust Management. Procedia Computer Science **220**, 405–412 (2023)
2. Rjoub, G., Wahab, O.A., Bentahar, J., Cohen, R., Bataineh, A.S.: Trust-augmented deep reinforcement learning for federated learning client selection. Information Systems Frontiers **26**(4), 1–18 (2022)
3. Haseeb, K., Saba, T., Rehman, A., Ahmed, Z., Song, H.H., Wang, H.H.: Trust management with fault-tolerant supervised routing for smart cities using the internet of Things. IEEE Internet Things J. **9**(22), 22608–22617 (2022)
4. Din, I.U., Awan, K.A., Almogren, A., Kim, B.S.: ShareTrust: Centralized trust management mechanism for trustworthy resource sharing in industrial Internet of Things. Comput. Electr. Eng. **100**, 108013 (2022)
5. Magdich, R., Jemal, H., Ayed, M.B.: A resilient Trust Management framework towards trust related attacks in the Social Internet of Things. Comput. Commun. **191**, 92–107 (2022)
6. Awan, K.A., Din, I.U., Zareei, M., Almogren, A., Seo-Kim, B., Díaz, J.A.P.: Securing IoT with Deep Federated Learning: A Trust-based Malicious Node Identification Approach. IEEE Access **11**, 58901–58914 (2023)

7. Rjoub, G., Wahab, O.A., Bentahar, J., Bataineh, A.: Trust-driven reinforcement selection strategy for federated learning on IoT devices. Computing **106**(4), 1–23 (2022)
8. Mishra, S., Tyagi, A.K.: The role of machine learning techniques in the internet of things-based cloud applications. In: Artificial intelligence-based Internet of things systems, pp.105–135 (2022)
9. Arshad, Q.U.A., Khan, W.Z., Azam, F., Khan, M.K., Yu, H., Zikria, Y.B.: Blockchain-based decentralized trust management in IoT: systems, requirements, and challenges. Complex & Intelligent Systems, pp.1–22 (2023)
10. Rjoub, G., Bentahar, J. and Wahab, O.A.: Explainable Trust-aware Selection of Autonomous Vehicles Using LIME for One-Shot Federated Learning. In: *2023 International Wireless Communications and Mobile Computing (IWCMC)*, pp. 524–529. IEEE 2023
11. Ahmadian, M., Ahmadi, M., Ahmadian, S.: A reliable deep representation learning to improve trust-aware recommendation systems. Expert Syst. Appl. **197**, 116697 (2022)
12. Zarei, M.R., Moosavi, M.R., Elahi, M.: Adaptive trust-aware collaborative filtering for cold start recommendation. Behaviormetrika **50**(2), 541–562 (2023)
13. https://data.world/datasets/trust
14. Thirupathaiah, K., Qasymeh, M.: Optical Ultra-Wideband Nano-Plasmonic Bandpass Filter Based on Gap-Coupled Square Ring Resonators. IEEE Access **11**, 106095–106102 (2023)
15. Zhong, Z., Li, C.T., Pang, J.: Hierarchical message-passing graph neural networks. Data Min. Knowl. Disc. **37**(1), 381–408 (2023)

Author Index

A
Abbas, Safa F.   I-207
Agarwal, Tara Rani   II-39
Alam, Bashir   II-87
Anil Kumar, Pagidirayi   I-173
Anuradha, B.   I-173
Arzoo   II-260

B
Bahadure, Nilesh Bhaskarrao   II-271
Bal, A.   II-77
Batth, Jaspreet Singh   I-108
Ben, Denny   II-3
Bhatia, Anupam   I-257
Bhattacharya, R.   II-77
Bhatti, Emmy   I-57
Bisht, Vaasu   I-17
Bodla, Ashwin Kumar   I-124
Bonam, Abhishek Reddy   I-87

C
Chowdary, K. Kotaiah   I-74

D
Damodara, Rama Krishna   I-124
Deepa, R.   I-3
Dendhi, Prithika Reddy   I-87
Devi, K.   I-97
Dhara, R.   II-77
Dhinakar, P.   I-159
Dockara, Tirupathi Rao   II-154
Dodda, Ratnam   I-87
Dugar, Vansh   I-17

E
Elangovan, Muniyandy   I-97

G
Gangwar, Rakesh Chandra   II-176
Garg, Neha   I-97, II-105

Gautam, Nitin Ravi   II-230
Gokulapriya, R.   II-66
Goyal, Nidhi   II-53
Gupta, Swati   II-145

H
Hamid, Muhammad   II-87
Harishwaran, S.   II-66
Harshith, H. D.   I-192

I
Indoria, Devadutta   II-105

J
Jadhav, Rahul Namdeo   I-45
Jain, Shreyans   II-191
Jain, Shubbham   II-191
Jain, Tanmai   I-192
Jaswal, Shivani   II-130
Joy, Jisha   II-130

K
Kamthania, Deepali   II-191
Karthik, B. N.   II-105
Kashyap, Neeti   II-25
Kashyap, Seema   I-148
Kaur, Amanpreet   II-16
Kaur, Kiranbir   II-260
Kaur, Kiranjeet   I-108
Kaur, Prabhpreet   I-57
Kirthivasan, Pradeep Rajagopal   II-154
Krishan, Ram   II-204
Kumar, Atul   I-216
Kumar, Vivek   II-204
Kushwah, Rashmi   I-245

L
Lahoti, Anand   I-17
Latha, K. N. Madhavi   I-74
Lehal, Gurpreet Singh   I-216

M
Malhotra, Manisha II-116
Malik, Annu I-245
Mishra, Abhilash II-116
Mishra, Atul II-293
Mishra, Harsh II-230
Mohanty, Jaikishan I-135

N
Naim, Iram I-148
Neeraj, S. I-257
Neha, B. N. I-192

P
Pal, Om II-87
Pal, P. II-77
Pandey, Pragya I-34
Pathak, Krishna Kant II-3
Patni, Jagdish Chandra II-271
Pavan Sai, CH. I-233
Prabha, R. I-3
Prabu, R. I-159

R
Rajput, Aarav I-17
Rajput, Sudhir Kumar II-271
Rakheja, Pankaj II-25
Ranjan, Arti II-246
Rasheed, Areeg Fahad I-207
Ravinder, M. II-246
Rubi II-282

S
Sahu, Dinesh Prasad II-282
Sakinala, Srinidhi I-87
Sambandam, Rakoth Kandan II-66
Saurabh, Sonal II-145
Saxena, Sandeep II-230
Senthil Kumar, C. I-273
Senthil, G. A. I-3
Sethia, Divyashikha I-135, II-217
Shaik, Nazeer I-97, II-105, II-282
Sharma, Abha I-17, II-3
Sharma, Priyanka I-34
Shilpa II-16
Shimona, S. I-3
Shrivastav, A. K. II-282
Shukla, Arvind Kumar I-148
Sidharth Reddy, B. I-233
Singh, Aashdeep II-16
Singh, Arun Kumar II-230
Singh, Gurpreet II-16
Singh, Jagendra I-97, II-105, II-282
Singh, Konjengbam Roshan II-217
Singh, Roohi II-176
Singh, Sadhana I-34
Soni, Sagar II-191
Sri Chandana Charudatta, D. I-233
Sridevi, S. I-3
Srinivas, S. I-233
Sudhagar, G. I-45

T
Thakur, Nikita II-39
Tiwari, Mohit I-97, II-105, II-282

V
Vijay Anand, R. I-273
Vimala Devi, J. I-192

Y
Yadav, Gagan S. I-192
Yadav, Monika II-293

Z
Zarkoosh, M. I-207

Printed in the United States
by Baker & Taylor Publisher Services